George Chandler Whipple

The Microscopy of Drinking-Water

George Chandler Whipple

The Microscopy of Drinking-Water

ISBN/EAN: 9783743313439

Manufactured in Europe, USA, Canada, Australia, Japa

Cover: Foto ©berggeist007 / pixelio.de

Manufactured and distributed by brebook publishing software (www.brebook.com)

George Chandler Whipple

The Microscopy of Drinking-Water

THE MICROSCOPY
OF
DRINKING-WATER.

BY

GEORGE CHANDLER WHIPPLE,

Biologist and Director of Mt. Prospect Laboratory, Department of Water Supply, Brooklyn, N. Y.; formerly Biologist of the Boston Water Works.

FIRST EDITION.

FIRST THOUSAND.

NEW YORK:
JOHN WILEY & SONS.
London: CHAPMAN & HALL, Limited.
1899.

DEDICATED
TO
My Father.

PREFACE.

THIS book has a twofold purpose. It is intended primarily to serve as a guide to the water-analyst and the water-works engineer, describing the methods of microscopical examination, assisting in the identification of the common microscopic organisms found in drinking-water and interpreting the results in the light of environmental studies. Its second purpose is to stimulate a greater interest in the study of microscopic aquatic life and general limnology from the practical and economic standpoint.

The work is elementary in character. Principles are stated and briefly illustrated, but no attempt is made to present even a summary of the great mass of data that has accumulated upon the subject during the last decade. The illustrations have been drawn largely from biological researches made at the laboratory of the Boston Water Works and from the reports of the Massachusetts State Board of Health. In considering them one should remember that the environmental conditions of the Massachusetts water-supplies are not universal, and that every water-supply must be studied from the standpoint of its own surroundings. So far as the microscopic organisms are concerned, however, the troubles that they have caused in Massachusetts may be considered as typical of those experienced elsewhere.

The descriptions of the organisms in Part II are necessarily brief and limited in number. The organisms chosen for description are those that are most common in the water-supplies of New England, and those that best illustrate the most important groups of microscopic animals and plants. In many cases whole families and even orders have been omitted, and some readers will doubtless look in vain for organisms that to them seem important. The omissions have been made advisedly and with the purpose of bringing the field of microscopic aquatic life within the scope of a practical and elementary survey. For the same reason the descriptions stop at the genus and no attempt has been made to describe species and varieties. Notwithstanding this it is believed that the illustrations and descriptions are complete enough to enable the general reader to obtain a true conception of the nature of the microscopic life in drinking-water and to appreciate its practical importance. To the student they must serve as a skeleton outline upon which to base more detailed study.

The illustrations, for the greater part, have been drawn from living specimens or from photo-micrographs of living specimens, but some of them have been reproduced from published works of standard authority. Among these may be mentioned: Pelletan and Wollé on the Diatomaceæ; Wollé, Rabenhorst, and Cooke on the Chlorophyceæ and Cyanophyceæ; Zopf on the Fungi; Leidy, Bütschli, and Kent on the Protozoa; Hudson and Goss on the Rotifera; Baird and Herrick on the Crustacea; Lankester on the Bryozoa; Potts on the Spongidæ; and Griffith and Henry on miscellaneous organisms.

PREFACE.

This book has been prepared during the leisure moments of a busy year. Its completion has been made possible by the kind assistance of my present and former associates in the laboratories of the Boston and Brooklyn water-supply departments and of other esteemed friends, to all of whom I tender my sincere thanks. I desire also to acknowledge the valuable assistance of my wife, Mary R. Whipple, in revising the manuscript and correcting the proof. To many others I am indebted indirectly, and among them I cannot refrain from mentioning the names of Prof. Wm. T. Sedgwick of the Massachusetts Institute of Technology; Mr. Geo. W. Rafter, C.E., of Rochester, N. Y.; and Mr. Desmond FitzGerald, C.E., formerly Superintendent of the Boston Water Works and now Engineer of the Sudbury Department of the Metropolitan Water Works. To Prof. Sedgwick and Mr. Rafter water-analysts are indebted for the most satisfactory practical method of microscopical examination of drinking-water yet devised, and Mr. FitzGerald will be remembered not only as an eminent engineer but as the founder and patron of the first municipal laboratory for biological water-analysis in this country.

GEORGE CHANDLER WHIPPLE.

NEW YORK, January, 1899.

CONTENTS.

PART I.
CHAPTER I.
HISTORICAL.

PAGE

Early Investigators in Europe and the United States.—Cloth Method.—Kean's Sand Method.—Sedgwick's Improvements.—Sedgwick-Rafter Method.—Recent Improvements.—Plankton Studies in Europe and America.................................... 1

CHAPTER II.
THE OBJECT OF THE MICROSCOPICAL EXAMINATION.

Sanitary Analyses.—Interpretation of Analyses.—Use of Microscopical Examination in Indicating Sewage Contamination—in Explaining the Chemical Analyses—in Explaining the Turbidity and Odor of Waters—in Studying the Food of Fishes........................ 8

CHAPTER III.
METHODS OF MICROSCOPICAL EXAMINATION.

Sedgwick-Rafter Method.—The Filter.—Concentration.—The Cell.—The Microscope.—Enumeration.—Sources of Error.—Precision of the Method.—Results of Examination.—The Standard Unit.—Records.—The Plankton Net Method.—The Plankton Pump.—The Planktonokrit.. 15

CHAPTER IV.
MICROSCOPIC ORGANISMS IN WATER FROM DIFFERENT SOURCES.

Rain-water.—Ground-water.—River-water.—Canals.—Raphidomonas in the Lynn Water.—Pond-water............................. 41

CHAPTER V.
LIMNOLOGY.

Physical Properties of Water.—Compressibility.—Density.—Mobility.—Thermal Stratification.—Diathermancy.—Temperature of Lakes.—Methods of Observation.—Thermophone.—Seasonal Variation of Temperature.—Periods of Circulation.—Periods of Stagnation.—Thermocline.—Classification of Lakes according to Temperature.—Transmission of Light by Water.—Color of Water.—Seasonal Change of Color.—Bleaching of Color by Sunlight.—Turbidity of Water.—Transparency of Water.—Absorption of Light by Water... 50

CHAPTER VI.
GEOGRAPHICAL DISTRIBUTION OF MICROSCOPIC ORGANISMS.

Common Organisms Classified according to the Frequency of Their Occurrence.—Statistics of Their Occurrence in Massachusetts Surface-water Supplies.—Relation of Each Class of Organisms to the Sanitary Chemical Analyses 77

CHAPTER VII.
SEASONAL DISTRIBUTION OF MICROSCOPIC ORGANISMS.

Seasonal Succession of Organisms.—Spring and Autumnal Growths of Diatoms.—Explanation of this Seasonal Distribution.—Effect of Temperature.—Effect of Light.—Heliotropism.—Food-material.—Stagnation.—Seasonal Distribution of Chlorophyceæ, Cyanophyceæ, Schizophyceæ, Fungi, Protozoa, Rotifera, Crustacea..... 93

CHAPTER VIII.
HORIZONTAL AND VERTICAL DISTRIBUTION OF MICROSCOPIC ORGANISMS.

Littoral Organisms.—Limnetic Organisms.—Effect of Winds and Currents on Horizontal Distribution.—Conditions Affecting Vertical Distribution.—Growth above the Thermocline.—Effect of the Specific Gravity of Organisms.—Peculiar Vertical Distribution of Mallomonas.—Protozoa.—Statistics of Vertical Distribution...... 105

CHAPTER IX.
ODORS IN WATER-SUPPLIES.

The Senses of Taste and Odor.—Odors Caused by Organic Matter.—Odors of Decomposition.—Odors Caused by Living Organisms.—Character of Odoriferous Substances.—Intensity of Odors.—Characteristic Odors of Different Organisms.—Extent to which

Water-supplies are Afflicted with Odors.—Cucumber Odor Not Caused by Fresh-water Sponge.—Cucumber Odor in Boston Water.. 113

CHAPTER X.
STORAGE OF SURFACE-WATER.

Clean Watersheds.—Effect of Swamps.—Anabæna in Cedar Swamp.—Drainage of Swamps.—Self-draining Watersheds.—Self-draining Reservoirs.—Stagnation of Water in Deep Reservoirs.—Lake Cochituate.—Removal of Organic Matter from the Sides and Bottom of Reservoirs.—Blow-off at the Bottom of Deep Reservoirs 150

CHAPTER XI.
STORAGE OF GROUND-WATER.

Ground-water to be Stored in the Dark.—Growth of Organisms in Open Reservoirs.—Storage of Surface-water and Ground-water Together.—Asterionella in Brooklyn Water-supply.—Storage of Impure Ground-water.—Crenothrix.—Storage of Filtered Water.. 141

CHAPTER XII.
GROWTH OF ORGANISMS IN WATER-PIPES.

Effect of Pipes upon the Biology of Water.—Temperature.—Microscopic Organisms.—Amorphous Matter.—Bacteria.—Effect of Water upon the Biology of Water-pipes.—Hamburg.—Rotterdam.—Boston.—Food of Organisms Dwelling in Pipes.—Polyzoa.—Fresh-water Sponge.—Effect of Pipe-moss.—Friction.—Odor of Water.—Paludicella in Brooklyn................................. 146

PART II.

CHAPTER XIII.
CLASSIFICATION OF MICROSCOPIC ORGANISMS.

Table of Classification.. 156

CHAPTER XIV.
DIATOMACEÆ.

Diatom Cells.—Shape and Size.—Markings.—Cell-contents.—External Secretions.—Movement.—Multiplication.—Reproduction.—Classification.—Description of Genera................................. 157

CHAPTER XV.
SCHIZOMYCETES.

Schizophyceæ.—Schizomycetes.—Characteristics.—Description of Genera.. 176

CHAPTER XVI.
CYANOPHYCEÆ.

Characteristics.—Description of Genera............................ 179

CHAPTER XVII.
CHLOROPHYCEÆ.

Algæ.—Chlorophyceæ.—Characteristics.—Description of Genera...... 188

CHAPTER XVIII.
FUNGI.

Characteristics.—Description of Genera............................ 205

CHAPTER XIX.
PROTOZOA.

General Characteristics.—The Protozoan Cell.—Rhizopoda.—Mastigophora.—Infusoria.—Description of Genera.................... 209

CHAPTER XX.
ROTIFERA.

Characteristics.—Description of Genera............................ 232

CHAPTER XXI.
CRUSTACEA.

Characteristics.—Description of Genera............................ 240

CHAPTER XXII.
BRYOZOA (POLYZOA).

Characteristics.—Description of Genera............................ 245

CHAPTER XXIII.
SPONGIDÆ.

Characteristics.—Description of Genera............................ 248

CHAPTER XXIV.
MISCELLANEOUS ORGANISMS.

Aquatic Plants.—Aquatic Animals 251

APPENDIX A. COLLECTION OF SAMPLES............................... 253
" B. TABLES AND FORMULÆ................................ 256
" C. BIBLIOGRAPHY 260
" D. GLOSSARY TO PART II............................... 290
INDEX.. 293

THE MICROSCOPY OF DRINKING-WATER.

PART I.

CHAPTER I.

HISTORICAL.

THE study of the microscopic organisms in water dates back to the seventeenth century. With the invention of the compound microscope enthusiastic observers began to search ponds and streams and ditches for new and varied kinds of microscopic life. Among the pioneers in this field of Natural History were Hooke (1665), Leeuwenhoek (1675), Ray (1724), Hudson (1762), Müller (1773), Dillwyn (1809), Kützing (1834), Ehrenberg (1836), Dujardin (1841), and Stein (1849).

It was not until 1850 that the study of the organisms in drinking-water was recognized as having a practical sanitary value. Dr. Hassall of London was the first to call attention to it. His method of procedure is unknown, but in all probability it consisted of the examination of a few drops of the sediment collected in a deep vessel after allowing the water to stand for a longer or shorter interval. Radlkofer (1865) of Munich, and Cohn (1870), Hirt (1879) and Hulwa of Breslau, pursued the study and emphasized its importance, but they made no radical improvement in the method.

In 1875 Dr. J. D. Macdonald of London suggested improvements in the sedimentation method, and made a rude attempt to obtain quantitative results by allowing the water to settle for a definite length of time, collecting the sediment on a removable glass disk or watch-glass at the bottom of a tall jar, and afterwards transferring this glass disk with its accumulated sediment to the stage of the microscope for direct examination.

In 1884 Dr. H. C. Sorby of England attempted to obtain a more exact enumeration by passing a gallon of the sample through a fine sieve (200 meshes to an inch) and then washing the collected organisms into a dish and in some way counting then.

In America important researches were made by Torrey, Vorce, Mills, Leeds, Potts, Nichols, Farlow, and others, but previous to 1888 the work was chiefly of a qualitative character.

In 1887 the Massachusetts State Board of Health began a systematic examination of all the water-supplies of the State, and two years later the State Board of Health of Connecticut began a similar but less extensive series of examinations. In 1889 the Water Board of the City of Boston established a biological laboratory * at the Chestnut Hill Reservoir for the purpose of studying systematically the biological character of the various sources of supply. In 1893 a small laboratory was established by the Public Water Board of the City of Lynn, Mass. In 1897 Mt. Prospect Laboratory, connected with the Department of Water Supply of Brooklyn, N. Y., was equipped and put in operation. It is devoted to general water analysis, and the microscopical examination of water

---

* For the first eight years of its existence it was conducted by the author under the general direction of Mr. Desmond FitzGerald, Superintendent of the Western Division of the Water Works.

from the different sources of supply forms an important part of the routine work.

Similar biological work has been lately undertaken by health departments and by water experts in other parts of the country.

The reports of the various laboratories show that during the last ten years more than fifty thousand samples of water have been submitted to microscopical examination in New England and New York, and that this number is increasing at the rate of about eight thousand a year.

The method of microscopical examination first used by the Massachusetts State Board of Health was that suggested by Mr. G. H. Parker. A piece of cotton cloth was tied firmly over the end of a glass funnel and 200 c.c. of the sample were made to pass through it. The organisms were left as a deposit on the cloth. After this straining the cloth was removed and inverted over an ordinary microscopical slip. The organisms, together with a small quantity of water, were dislodged upon the slip by blowing downwards upon the cloth through a piece of glass tube. This method was useful, but it did not give accurate quantitative results. Mr. F. F. Forbes of Brookline, Mass., used a modification of the cloth method. The water was filtered as in Parker's method, but the neck of the funnel passed into a tank from which the air was exhausted by an aspirator. This hastened the filtration and allowed a larger amount of water to be filtered.

The present method of examination was foreshadowed in the work of Mr. A. L. Kean. He filtered 100 c.c. of his samples through a small quantity of coarse sand placed at the bottom of a glass funnel and supported by a plug of wire gauze. After filtration the plug was removed and the sand with its contained organisms was washed into a watch-glass with 1 c.c. of water. This was stirred up to separate the organisms from the sand and a portion was transferred to a cell holding one cubic millimeter. From the number of organisms found in this cell the approximate number originally present in the water could be obtained. This method has become known as the "sand method."

In 1889 Prof. Wm. T. Sedgwick and Mr. Geo. W. Rafter made valuable improvements upon Kean's original idea. Prof. Sedgwick suggested the use of a cell much larger than that used by Kean, bounded by a brass rim and having an area of 1000 square millimeters ruled by a dividing engine into 1000 squares. The filtration was made as before, and the sand was washed into the cell with one or two cubic centimeters of water and distributed over the bottom. The cell was then placed under the microscope and the organisms counted in a certain number of the small squares. From this count the number of organisms present in the sample was estimated. A modification of this method was the one first used by the Connecticut State Board of Health. In the Connecticut method precipitated silica was used instead of sand for the filtering medium, and this was supported upon a pledget of absorbent cotton.

Mr. Rafter's improvements consisted in the substitution of a ruled square in the ocular of the microscope for the ruling upon the plate, in the separation of the sand from the organisms by decantation, in the use of a cell covered by a cover-glass and containing just one cubic centimeter, and in the use of a specially constructed mechanical stage. The Sedgwick-Rafter method has been modified somewhat by recent experimenters,* but its essential character has not been changed.

---

* Dr. Gary N. Calkins substituted a perforated rubber stopper capped

While sanitarians have been pursuing the study of the microscopic organisms because of their effect on the quality of water-supplies, other scientists have approached the subject from an entirely different standpoint. In the same year that the Massachusetts State Board of Health began its examination of the water-supplies of the State, Victor Hensen of the University of Kiel, Germany, published a description of a new method of studying the minute floating organisms found in lakes. To these organisms he gave the name "plankton," * a collective word applied to all minute animals and plants that float free in the water and that are drifted about by waves and currents. Plants attached to the shore, and animals that possess strong powers of locomotion, are not included in the plankton, but fragments of shore plants, fish-eggs, young fish-fry, etc., are included. The term "plankton," however, may be said to be synonymous with the term "microscopic organisms" of the sanitary biologist.

Hensen's method is radically different from the Sedgwick-Rafter method. The latter is strictly a laboratory process. The samples of water operated on are small; the concentration of the organisms is made in the laboratory. Hensen devised a net by which the organisms could be concentrated in the field, so that only the collected material need be taken to the laboratory.

Even before the publication of Hensen's paper, scientists

---

by a circle of bolting-cloth in place of the plug of wire gauze. Mr. D. D. Jackson suggested a cylindrical funnel in place of the ordinary flaring chemical funnel, and added an attachment at the lower end to control the concentration and prevent the sand from becoming dry. The author has graduated the funnels, designed a simple automatic concentrating device, and applied an aspirator to hasten the filtration. He also designed the ocular micrometer and the record blank now used, and suggested the idea of a standard unit of size for estimating the organisms and amorphous matter.

* From the Greek *planktos*, wandering.

on the Continent had become interested in the study of lakes. The early observations of Prof. F. A. Forel, of Morges, Switzerland, on Lake Geneva were followed by the establishment of a Limnological Commission in Switzerland. Under its direction many valuable lines of physical and biological research were undertaken. This was followed in 1890 by an International Commission. From this time increased attention has been given to the biology of ponds and lakes. A biological station was established by Zacharias at Lake Plön in 1891, and a group of scientists have contributed important articles to its annual reports. Apstein at Kiel, Schroeter at Zurich, and many others have made extensive and valuable observations. Biological stations are multiplying and the work is being extended to France, Italy, Austria, Denmark, and Norway.

Similar investigations have been carried on in the United States. In 1893 Prof. J. E. Reighard, acting under the direction of the Michigan Fish Commission, made a biological study of Lake St. Clair. This was followed by an examination of Lake Michigan by Prof. Henry B. Ward, and by studies of the crustacea in Lake Mendota by Prof. E. A. Birge, and in Green Lake by Prof. C. Dwight Marsh.

Biological stations have been recently established by a number of Western universities on and in the vicinity of the Great Lakes.

For several years Prof. James I. Peck, acting under the direction of the U. S. Fish Commission, has been making important studies of the food of certain fishes, notably the menhaden. He uses the Sedgwick-Rafter method instead of the plankton net for concentrating the microscopic organisms.

In 1896 Dr. C. S. Dolley of Philadelphia suggested the use of the centrifugal machine for the purpose of concentrating the microscopic organisms. This "planktonokrit," as it is called, has not been developed to completeness, but recent experiments by Field, Kofoid, and others show that it is likely to prove a valuable adjunct to the Sedgwick-Rafter method, but it is not likely to supplant it.

Prof. H. B. Ward and Mr. Chas. Fordyce have recently devised a "plankton pump" for collecting crustacea and other plankton organisms at particular depths below the surface of a lake. It seems to be a decided improvement over the plankton net.

CHAPTER II.

THE OBJECT OF THE MICROSCOPICAL EXAMINATION.

A COMPLETE sanitary examination of water, as conducted in modern laboratories, consists of three parts,—the physical, the biological, and the chemical. The biological examination may be divided into the microscopical and the bacterial, because of different methods employed. The data obtained are as follows:

PHYSICAL EXAMINATION.
>Temperature — Turbidity — Sediment — Color—Odor, both cold and hot.

BIOLOGICAL EXAMINATION.
Microscopical.
>Number of microscopic organisms per c.c.—Amount of inorganic, amorphous matter, etc.

Bacteriological.
>Number of bacteria per c.c.—Presence of intestinal or pathogenic bacteria.

CHEMICAL EXAMINATION.
>Total Residue on Evaporation—Loss on Ignition—Fixed Solids — Alkalinity — Hardness — Chlorine — Iron—Nitrogen as Albuminoid Ammonia—Nitrogen as Free Ammonia—Nitrogen as Nitrites—Nitrogen as Nitrates — Total Organic Nitrogen (Kjeldahl Method)—Oxygen consumed—Dissolved Oxygen. (The last three are of use only in special cases.)

Such an analysis is intended to show whether or not the water is of such a character that it would cause sickness if used for drinking; whether or not it contains anything that would render it distasteful or unpalatable; and whether or not it contains any ingredient that would make it unfit for laundry use or for general domestic or industrial purposes. Sanitary examinations are necessary also in studying the effect of processes of purification.

Opinions regarding the function and value of sanitary water analyses have undergone a change in recent years. The numerical results of a single analysis of a sample of water, when considered by themselves, are now believed to have little intrinsic value. It has been found that the value of the analysis lies in its interpretation, and that each part of the analysis must be interpreted by comparison with all the other parts and in the light of exact knowledge of the environment of the water. The interpretation of an analysis is as much a matter of expert skill as is the making of the analysis itself. The physical, biological, and chemical examinations should be interlocking in their testimony, yet these different parts are to be given different weight in the study of different problems. For example, in the detection of pollution the chemical and bacterial examinations furnish the most information, in the study of the æsthetic qualities of a water the physical and microscopical examinations are most important, while in investigations concerning the value of a water for industrial purposes the chemical and physical examinations sometimes suffice.

The biological examination is concerned with the micro-organisms found in water. The term "micro-organisms," when used in its broadest and most literal significance, includes all organisms which are invisible or barely visible to the naked eye. It is frequently used in a narrower sense, however, as a synonym for bacteria. Using the word in its broad sense we may divide the micro-organisms found in water into two classes, as suggested by Prof. Sedgwick.

MICRO-ORGANISMS. Organisms, either plants or animals, invisible or barely visible to the naked eye.
{
Microscopic Organisms.
Not requiring special culture.
Easily studied with the microscope.
Microscopic in size, or slightly larger.
Plants or animals.

*Bacterial Organisms.**
Requiring special cultures.
Difficultly studied with the microscope.
Microscopic or sub-microscopic in size.
Plants.
}

This subdivision is convenient for the sanitarian as well as for the biologist, because the two classes of organisms affect water in different ways. With certain reservations it may be said that the bacteria make a water unsafe, the microscopic organisms make it unsavory.

Microscopical Examination.—The microscopical examination of water may be considered in four aspects: 1. As indicating sewage contamination. 2. As explaining the chemical analysis. 3. As explaining the cause of turbidity, odors, etc., in water. 4. As a method of studying the food of fishes and other aquatic animals.

1. The microscopical examination cannot be depended upon to determine the pathogenic qualities of a drinking-

---

* The bacteria are not considered in this volume. The reader is referred to the numerous works on Bacteriology, and in particular to "Micro-organisms in Water," by Percy and G. C. Frankland.

water. To be sure, the germs of disease are microscopic bodies, and when artificially cultivated or when found in the tissues of the body they can be studied with microscopes of high power. But when scattered through a mass of water they cannot be detected by ordinary microscopical methods, because of their small size and because they are greatly outnumbered by the ordinary water bacteria. It is questionable whether they can be discovered even by methods of culture. Not only may water contain pathogenic bacteria without discovery, it may contain the ova or larvæ of some of the endoparasites of man. It is probable that endoparasitic diseases are more common than has been generally supposed; and while diseased pork, beef, etc., are the chief agencies of infection, it is known that water polluted by animal excrement may contain the ova or larvæ of such endoparasites as *Tænia solium*, *Tænia saginata*, *Bothriocephalus latus*, *Ascaris lumbricoides*, *Trichocephalus dispar*, and *Anchylostomum duodenale*. Infection of animals by the drinking of water contaminated by barnyard wastes has been several times recorded, while a microscopical examination of the water has seldom revealed the presence of the suspected ova or larvæ. This is not because they are too minute to be detected, but because the quantity of water examined is necessarily too small.

The microscopical examination cannot show definitely whether a water is polluted by sewage unless the pollution is excessive. It can, however, give evidence which, taken with the chemical and bacterial examinations, may establish the proof. A microscopical examination of sewage reveals few of the living organisms that are found ordinarily in water. Ciliated infusoria, such as Paramæcium and Trachelocerca; fungus forms, such as mold hyphæ, Saprolegnia, Leptomitus, Leptothrix, and Beggiatoa; and miscellaneous objects, such as yeast-cells, starch-grains, fibres of wood and paper, fibres of muscle, epithelial cells, threads of silk, woolen, cotton and linen, insect scales, feather barbs, etc., may be observed. Most of these objects are foreign to unpolluted water, and their presence in a sample of water leads one to suspect its purity.

Furthermore, there are other organisms, such as *Euglena viridis*, which live on decaying vegetable matter and which, though not found in sewage, are often associated with it in polluted water. Their presence in a sample is a cause of suspicion. These evidences, however, should be weighed only in connection with an environmental study and with the entire sanitary analysis. The *common* microscopic organisms found in water are not themselves the cause of disease, nor does their presence indicate sewage pollution.

2. The chemical examination determines the amount of organic matter that a sample of water contains, but it does not determine the nature of it. As the character and condition of the organic matter is very important from the sanitary point of view, the microscopical examination gives valuable information by showing not only whether the organic matter in suspension is vegetable or animal, but by determining whether it is made up of living organisms or of decomposing fragments. For example, the amount of albuminoid ammonia in suspension is sometimes so great that one might suspect that the water was polluted did the microscope not show that the high figure was due to a growth of some organism. Or in a series of samples from a reservoir it might be difficult to account for a sudden decrease in the nitrates or free ammonia were it not for the appearance of some microscopic organism that had appropriated the nitrogen as a part of its food.

3. By far the most important service that the microscopical examination renders is that of explaining the cause of the taste and odor of a water and of its color, turbidity, and sediment. Several of the microscopic organisms give rise to objectionable odors in water and, when sufficiently abundant, have a marked influence on its color. They also make the water turbid and cause unsightly scums and sediments to form. Upon all such matters related to the æsthetic qualities of a water the microscopical examination is almost the only means of obtaining reliable information.

4. The microscopic organisms form the basis of the food-supply of fishes and other aquatic animals. Sometimes the relation is a direct one; that is, the microscopic organisms are themselves eaten by fish. This has been well illustrated by Prof. Peck. The menhaden swims with its mouth open, and is provided with a peculiar filtering apparatus by which the minute organisms are caught. It has been shown that the presence or absence of these fish from certain sections of the Massachusetts coast depends upon the abundance of microscopic life in the water, and also that the weight of fish of any particular length depends upon the quantity of this food material at hand.

The relationship between the plankton and fish life is not always so direct. In many cases the fish feed upon crustacea and insect larvæ; the crustacea feed upon the rotifera and protozoa; the rotifera and protozoa feed upon algæ; while the algæ nourish themselves by the absorption of soluble inorganic substances.

The interrelations between different organisms of the lower world, and between the organisms and their environment are matters of intense scientific interest, and limnology is fast assuming an important place in scientific literature.

The physical condition of lakes, the currents, waves, temperature, and transparency of water, the chemistry of water, the life-history of organisms, and various bio-chemical and bio-physical problems are more and more attracting the attention of scientists and of water-works engineers.

CHAPTER III.

METHODS OF MICROSCOPICAL EXAMINATION.

THE most important methods of microscopical examination of water now in use are: 1. The Sedgwick-Rafter Method; 2. The Plankton Net Method; 3. The Plankton Pump Method; 4. The Planktonokrit. They differ chiefly in the manner of concentrating the organisms.

I. THE SEDGWICK-RAFTER METHOD.

The Sedgwick-Rafter Method consists of the following processes: the filtration of a measured quantity of the sample through a layer of sand upon which the organisms are detained; the separation of the organisms from the sand by washing with a small measured quantity of filtered or distilled water and by decanting; the microscopical examination of a portion of the decanted fluid; the enumeration of the organisms found therein; and the calculation from this of the number of organisms in the sample of water examined. The essential parts of the apparatus are the filter, the decantation-tubes, the cell, and the microscope with an ocular micrometer.

The Filter.—The sand may be supported upon a plug of rolled wire gauze at the bottom of an ordinary glass funnel 7 or 8 inches in diameter, but the cylindrical funnel shown in Fig. 1 is preferable. The inside diameter of this funnel at the top is 2 inches; the distance from the top to the beginning of the slope is 9 inches; the length of the slope is about 3 inches; the length of the tube of small bore is 2½ inches, and its inside diameter is ½ inch. The capacity of the funnel is 500 c.c. The support for the sand consists of a perforated rubber stopper pressed tightly into the stem of the funnel and capped with a circle of fine silk bolting-cloth. The circles of bolting-cloth may be cut out with a wad-cutter. Their diameter should be a little less than that of the small end of the rubber stopper. When moist the cloth readily adheres to the stopper. The sand resting upon the platform thus prepared should have a depth of at least three fourths of an inch. The quality of the sand is important. Ordinary sand is unsatisfactory unless very thoroughly washed. Pure ground quartz is preferable. Its whiteness is a decided advantage. The necessary degree of fineness of the sand depends somewhat upon the character of the water to be filtered. A sand which will pass through a sieve having 60 meshes to an inch, but which will be retained by a sieve having 120 meshes, will be found satisfactory for most samples. Such a sand is described as a 60–120 sand. When very minute organisms are present a finer sand must be used,—say a 60–140 sand. The sand used for many years by the author had the following composition:

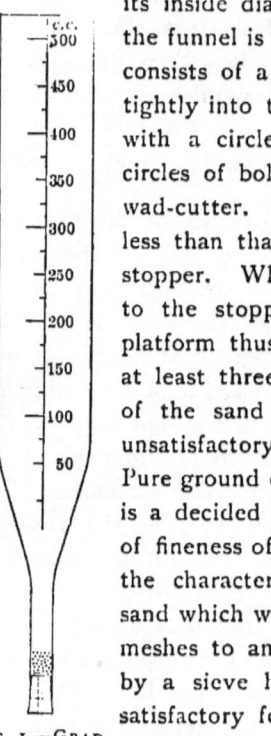

FIG. 1.—GRADUATED CYLINDRICAL FUNNEL USED IN THE SEDGWICK-RAFTER METHOD.

Size of sand-grains....	40–60	60–80	80–100	100–120	120–140	
Percentage by weight..	20	20	38	18	4	= 100

The filters may be arranged conveniently in a row against the laboratory wall as shown in Fig. 2. The filtered water

FIG. 2.—BATTERY OF FILTERS. SEDGWICK-RAFTER METHOD.

may be collected in a sloping trough and carried to a sink, or jars may be placed under the separate funnels. A hinged covering-shelf above the filters is useful to prevent the access of dust.

The sample to be filtered may be measured in a graduated cylinder or flask, or the filter-funnel itself may be graduated. The graduated filter-funnel is especially useful for field work, as it saves the necessity of carrying an additional graduate. The quantity of water that should be filtered depends upon the number of organisms and the amount of amorphous matter present. An inspection of the sample will enable one to judge the proper amount. Ordinarily 1000 c.c. for a ground-water and 500 c.c. for a surface-water will be found satisfactory. In some cases 250 c.c. or even 100 c.c. of a surface-water will be found more convenient. When the water is poured into the funnel care should be taken not to disturb the sand more than is necessary, otherwise organisms are liable to be forced through the filter. The best plan is to make the sand compact by pouring in enough distilled water to just about fill the neck of the funnel and to pour in the measured sample before the sand has become uncovered. The filtration ordinarily takes place in about half an hour, but occasionally a sample is so rich in organisms and amorphous matter that the filter becomes clogged. It then becomes necessary to agitate the sand with a glass rod or to apply a suction to hasten the filtration. If the filters are located near running water an aspirator may be attached to the faucet and connected with the filter by a rubber tube having a glass connection that fits the bore of the rubber stopper. The use of the aspirator enables the filtration to be made in a few minutes, and not only effects a saving in time, but reduces the error caused by the organisms settling on the sloping surface of the funnel.

Concentration.—As a result of the filtration the organisms and whatever other suspended matter the sample contained will have been collected on the sand. When all the water has passed through and before the sand has become dry the rubber stopper is removed and the sand with its accumulated organisms is washed down into a wide test-tube by a measured quantity of filtered or distilled water delivered from a pipette or an automatic burette. The amount of water used for washing depends upon the number of organisms collected on the sand. If 500 c.c. of the sample are filtered it is usually best to wash the sand with 5 c.c., thus concentrating the organisms one hundred times. The amount of water filtered divided by the amount of water used in washing the sand gives the "degree of concentration." The degree of concentration may vary from 10 to 500 according to the contents of the sample. Ordinarily it should be 50 or 100.

By shaking the test-tube the organisms will become detached from the sand-grains. If this is followed by a rapid decantation into a second test-tube most of the organisms, being lighter than the sand, will pass over with the decanted fluid, while the sand is left upon the walls of the first tube. To insure accuracy the sand should be washed a second time and the two decanted portions mixed together. If, for example, it is desired to concentrate a sample from 500 c.c. to 10 c.c. the sand should be washed twice with 5 c.c. and the two portions poured together. This will give a more accurate result than a single washing with 10 c.c.

Mr. D. D. Jackson has suggested the use of an attachment at the lower end of the funnel that automatically arrests the filtration as soon as the proper degree of concentration has been reached. This is illustrated in Fig. 3. The attachment fits over the rubber cork that supports the sand, uniting with the lower end of the neck of the filter by a ground joint. The

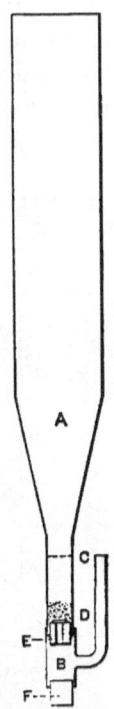

FIG. 3. — CONCENTRATING ATTACHMENT. SEDGWICK-RAFTER METHOD.

filtered water passes out of the open tube, and filtration stops as soon as the level of the water in the funnel has reached the elevation of the outlet-tube. This elevation is made such that the water remaining in the neck of the funnel is just sufficient to give the desired degree of concentration. It is necessary to allow for the volume of the sand and to use a definite amount of sand at each filtration. The 60–120 sand holds about 50% of water, and ordinarily about 2 c.c. of the sand are used; hence an allowance of 1 c.c. must be made for the water held in the sand. Thus, if it is desired to concentrate the organisms in 5 c.c. of water, the capacity from the bottom of the sand layer to the graduation on the stem of the funnel must be 6 c.c., i.e. 4 c.c. above the 2 c.c. of sand. After filtration the attachment is removed and a plug is inserted in the hole in the stopper to prevent loss of the concentrated fluid. The stopper then may be removed and the sand and organisms allowed to fall into a wide test-tube, after which the process is carried on as described below.

The same result has been accomplished by the author in a simpler way. In place of the attachment with the ground-glass joint a glass tube bent twice at right angles may be inserted in the opening in the rubber stopper as shown in Fig. 4. The funnel stem may be graduated and a series of bent tubes provided corresponding to different degrees of concentration.

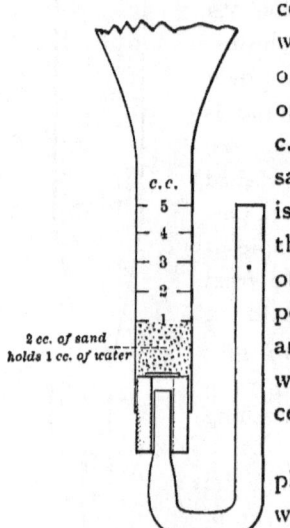

FIG. 4.—SIMPLE FORM OF CONCENTRATING ATTACHMENT. SEDGWICK-RAFTER METHOD.

If the operator is watching the filtration even this form of attachment is unnecessary, as the filtration may be stopped by inserting a plug in the rubber stopper as soon as the level of the water has fallen to the desired point. This method of concentrating is to be preferred to the usual one described above in which the surface of the sand is allowed to become uncovered before the sand is washed into the test-tube. As the use of either form of attachment described above retards the rate of filtration it is better not to put on the attachment until the water has fallen almost to the desired level.

If the concentrated water is allowed to stand in the funnel for any length of time some of the organisms are liable to become attached to the glass sides. To prevent error from this cause the neck of the funnel may be washed with a small measured quantity of filtered water, and this may be caught in the large test-tube and used for washing the sand a second time as described above.

The Cell.—The cell into which a measured portion of the concentrated fluid is placed for examination is made by cementing a rectangular brass rim to an ordinary glass slip. The internal dimensions of the cell are: length 50 mm., width 20 mm., and depth 1 mm. It therefore has an area of 1000 sq. mm. and a capacity of 1 c.c. A thick cover-glass (No. 3) having dimensions equal to those of the outside of the brass rim (55 mm. by 25 mm.) forms a roof to the cell. The concentrated organisms in the decantation-tube are distributed uniformly through the fluid by blowing into it through a pipette, and one cubic centimeter of the fluid is then transferred to the cell in such a manner as to distribute the organisms evenly over the entire area. This may be done by laying the cover-glass diagonally over the cell so that an opening is left at either end, and flowing the water in at one end while the air escapes at the other (see Fig. 5).

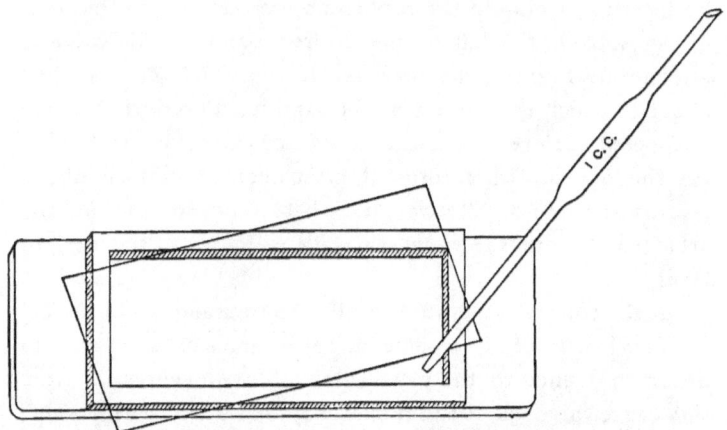

FIG. 5.—COUNTING-CELL, SHOWING METHOD OF FILLING. SEDGWICK-RAFTER METHOD.

The Microscope. — An expensive microscope is not needed for the numerical estimation of the common microscopic organisms found in water. A simple, compact stand with a $\frac{1}{2}$-inch objective and a 1-inch ocular is sufficient. For studying the organisms in detail and for general laboratory use in the study of water a large stand, with substage condenser, iris diaphragm, mechanical stage, etc., should be provided. The list of objectives should include a 2-inch, a $\frac{1}{2}$-inch, a $\frac{1}{4}$- or $\frac{1}{6}$-inch, and a $\frac{1}{12}$-inch homogeneous immersion, or their equivalents, and there should be several oculars magnifying from 4 to 12 times.

The ocular micrometer consists of a square ruled upon a thin glass disk which is placed upon the diaphragm of the ocular. The square is of such a size that with a certain combination of objective and ocular and with a certain tube-length of the microscope, the area covered by it on the stage is just one square millimeter. For convenience it should be subdivided as shown in Fig. 6. The size of the largest square is one square millimeter. The size of the smallest square is

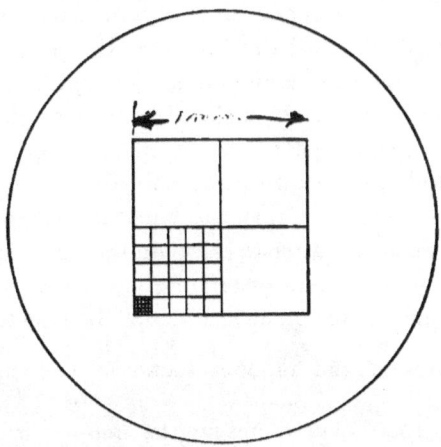

FIG. 6.—OCULAR MICROMETER USED IN THE SEDGWICK-RAFTER METHOD.

one standard unit.* The best micrometers are made by engraving, but a serviceable micrometer for occasional use may be made by photography.† With a ½-inch objective and a No. 3 ocular the square ruled for the ocular micrometer should be 7 mm. on a side. Before using the micrometer the proper tube-length must be ascertained by comparison with a stage micrometer.

Enumeration.—The cell, filled with the concentrated fluid, is placed upon the stage of the microscope and the organisms included within the area of the ruled square are counted. It is then moved so that another portion of the cell comes into the field of view and another square is counted.

---

* See page 29.

† This idea was suggested by Mr. Wallace Goold Levison, Brooklyn, N. Y.

This is continued until a sufficient number of representative squares has been examined. It is obviously impracticable to count all of the 1000 squares which compose the area of the cell. It is usually sufficient to count ten or twenty squares, but a larger number ought to be scrutinized. In counting the organisms it should be remembered that some are heavy and sink to the bottom, while others are light and rise to the top. The observer should make a practice of changing the focus of the microscope so that both the upper and lower portions of each square may be examined.

From the number of organisms found in the ten or twenty squares it is an easy matter to calculate the number originally present in one cubic centimeter of the sample. If t represents the number of organisms found in twenty squares, $\frac{t}{20}$ will represent the number found in one square, and $50t \left(= \frac{t}{20} \times 1000 \right)$ will represent the number in the entire cell, or in one cubic centimeter of the concentrated fluid. This divided by the degree of concentration will give the number of organisms in one cubic centimeter of the sample. For example, if the sample has been concentrated from 500 c.c. to 5 c.c. the degree of concentration will be 100, and therefore $\frac{50t}{100} = \frac{1}{2}t$ will represent the "number of organisms per c.c." in the sample. The figure by which the total number of organisms counted must be multiplied in order to reduce to "number per c.c." may be called the factor. Its value may be expressed as follows:

$$\text{Factor} = \frac{1000w}{nf},$$

in which f equals the number of cubic centimeters of water filtered; w, the number of cubic centimeters of water used in washing the sand; and n, the number of squares counted.

Sources of Error.—The operations of the Sedgwick-Rafter method involve several sources of error. They may be classified as follows:

1. Errors in sampling.
2. Funnel error, or the error caused by the organisms adhering to the sides of the funnel.
3. Sand error, or the error caused by imperfect filtration.
4. Error of disintegration, due to the breaking up of organisms on the surface of the sand.
5. Decantation error, or the error caused by the organisms adhering to the particles of sand, and by the water used in washing the sand being held back by capillarity during the process of decantation.
6. Errors caused by the organisms not being uniformly distributed in the cell.

Errors in Sampling.—These errors arise chiefly from the fact that organisms vary in specific gravity and in their behavior towards light. If the bottle containing the sample is allowed to stand even for a short time, some of the organisms will sink to the bottom, some will rise to the surface; some will collect on the side of the bottle towards the light, others will shun the light as much as possible; while some will attach themselves quite firmly to the sides of the glass. Evidently the bottle must be shaken before the portion for examination is withdrawn. Errors in sampling are common, but, to a great extent, are avoidable.

Funnel Error.—The funnel error, due to the organisms settling upon and adhering to the sloping sides of the funnel, varies greatly according to the character of the water filtered. It is highest in the case of samples rich in the Cyanophyceæ and amorphous matter. These, being of a somewhat gelatinous nature, adhere readily to the glass, making a rough surface on which other organisms lodge. If the funnel is wet when the sand is put in, some of the sand-grains are liable to adhere to the sloping walls. This tends to increase the deposition of organisms. The funnel error is less in the cylindrical funnels than in the flaring funnels. Slow filtration, whether due to the character of the funnel or to the sample filtered, increases the error,—indeed it may be said that the funnel error is almost proportional to the time of filtration. Numerically the funnel error may vary from 0 to 15%. A long series of experiments* on waters that varied greatly in character gave an average error of 1% for the organisms and 3% for the amorphous matter.

Sand Error.—The sand error, due to imperfect filtration, depends upon the character of the organisms, upon the size of the sand-grains, and upon the depth of the sand. In selecting a sand two opposing conditions must be adjusted. The sand must be fine enough to form an efficient filter, and yet the grains must be large enough to settle readily in the decantation-tubes. A $\frac{1}{2}$-inch layer of the sand described on page 16 ought not to give a sand error greater than 5% unless the water contains minute organisms. When very minute organisms are present in large numbers the error from incomplete filtration may be as great as 25% or even 50%. The effect of the size of the sand-grains on the sand error is well illustrated by the following table compiled from experiments by Calkins on the filtration of water containing yeast-cells and starch-grains:

Size of Sand.	Percentage Sand Error.	
	Yeast-cells.	Starch-grains.
40–60	21.6	4.4
60–80	8.7	7.3
80–100	5.3	7.4
100–120	3.3	1.2

* By the author.

Most of the organisms that pass through the sand do so during the early part of the filtration, before the sand has become compacted. If, before the sample is poured into the funnel, the sand is compacted by passing through it some distilled water, using the aspirator to increase the pressure, the sand error will be reduced considerably.

Errors of Disintegration. — Many of the microscopic organisms are extremely delicate. They are very susceptible to changed conditions of temperature, pressure, and light. As soon as a sample of water has been collected in a bottle some of the organisms begin to disintegrate; and if the sample stands long before examination and if it is submitted to the joltings of a long trip by express, some of the organisms will break up and become unrecognizable. The process of filtration helps to disintegrate them by bringing them in violent contact with the surface of the sand, but the method of concentrating the sample by arresting the filtration as described above reduces this error to a considerable extent by keeping the sand from becoming dry and by preventing many of the organisms from even reaching the surface of the sand. The errors due to disintegration during transit and before examination can be avoided only by making the examination at the time of collection. This is often necessary, particularly when one is searching for such delicate organisms as Uroglena. The errors of disintegration during filtration cannot be entirely avoided, but if the examination of the concentrated fluid is supplemented by a direct examination of the water gross mistakes may be prevented. Uroglena, Dinobryon, etc., may be detected in the sample with the naked eye after a little practice. They may be taken up with a pipette and transferred to the stage of the microscope. This direct examination is important and ought always to be made, but its value is qualitative and not quantitative.

Decantation Error.—The decantation error depends to a great extent upon care in manipulation. When the attempt is made to separate the organisms from the sand by agitating with distilled water in one test-tube and decanting into a second tube, some of the organisms remain behind attached to the sand-grains, and, what is quite as important, some of the water used in washing remains behind.

The two errors act in opposition. If the sand retains a larger percentage of organisms than of water, the figures in the result will be too low; if it retains a larger percentage of water than of organisms, the concentration will be too great and the figures in the result will be too high. With the fractional method of washing the sand and with due care in decanting the decantation error ought not to exceed 5 per cent.

Errors in the Cell.—The errors due to the unequal distribution of the organisms over the area of the cell are extremely variable and cannot be well stated in figures. If the concentrated fluid is evenly mixed and well distributed over the cell, if the count is made just as soon as the material in the cell has settled, and if a large number of squares are counted, the error will be reduced to a minimum. If a sample happens to contain such motile organisms as Trachelomonas they may collect at the edges of the cell in search of air, or if the cell stands in front of a window for any length of time organisms sensitive to light may migrate from one side of the cell to the other.

Precision of the Sedgwick-Rafter Method.—Examination of hundreds of samples has shown that the results are usually *precise* within 10%, i.e. two examinations of the same sample seldom differ by more than that amount. The accuracy, however, depends greatly upon the character of the organisms in the water examined.

Results of Examination—Standard Unit.—The microscopical examination of most samples of surface-water will show that the concentrated fluid contains minute organisms of various kinds, fragments of larger animals and plants, masses of a grayish or brownish flocculent material, and fine particles of inorganic matter. The inorganic or mineral matter is usually not considered in the Sedgwick-Rafter method; more information can be obtained by a direct examination of the sediment and by chemical analysis. The brownish flocculent material has been called "amorphous matter" because of its formless nature, and "zoogloea" because of its supposed bacterial origin. The term zoogloea has a definite meaning in bacteriology and is applied to a mass of bacteria held together by a more or less transparent glutinous substance. It is not strictly appropriate as applied to the brownish flocculent matter, which is not so much a collection of bacteria as the product of bacterial action. The word *phytogloea* might be used in its place, but the term "amorphous matter" is a broader term and quite as appropriate. The amorphous matter, then, includes all the irregular masses of unidentifiable organic matter. It does not include vegetable fibres, vegetable tissue, etc., nor does it include mineral matter except as this is intimately mixed with the flocculent material. The amorphous matter occurs in a finely divided state or in lumps of varying size. In order to correctly estimate its amount it is necessary to have some unit of size. A unit of volume is impracticable because of the great labor involved in determining the dimensions of the masses observed, but a unit of area approaches closely to what is desired. Such a unit was suggested by the author in 1889, and has come into use under the name of "standard unit." The standard unit is represented by the area of a square 20 microns* on a side, i.e. by 400 square microns.

The ocular micrometer shown in Fig. 6 was subdivided to correspond to this unit. The square, which covers one square millimeter on the stage of the microscope, is divided into four equal squares. Each of these quarters is subdivided into 25 smaller squares, and each of these squares contains 25 standard units. The eye will readily divide the side of a small square into fifths, and this division is the side of the standard unit square. If desired, one of the small squares may be further subdivided into squares the actual size of the standard unit as shown in the figure. This can be done on the micrometers made by photography, but not conveniently on those engraved.

The microscopic organisms vary in size and in their mode of occurrence. Some are found as separate individuals, some are joined together into filaments, or into masses or colonies; some are one-celled, some are many-celled; some are extremely simple, some are complex; some are scarcely larger than the bacteria, some are easily visible to the naked eye. It is difficult to establish a satisfactory system for counting these varied forms. If an individual count is adopted one has to decide what shall be the unit, whether a cell, or a filament, or a colony, or a mass. Practice has varied in this matter. The best system of counting by individuals is that used by the Massachusetts State Board of Health. All diatoms, desmids, rhizopods, crustacea, the unicellular algæ, and nearly all rotifera and infusoria are counted as individuals; the filamentous algæ are counted as filaments; the social forms of infusoria and rotifera are counted as colonies; and many of the algæ that occur as irregular thalli are counted as masses.

---

* One micron = .001 millimeter.

This system, which, for convenience, we may call the "individual counting system," does not always give satisfactory results. In the Boston water-supply it was found often that a sample which a simple inspection showed to be heavily laden with algæ and which was offensive both in appearance and in odor gave a low figure in the count, while a sample that was clear and agreeable to the taste gave a very high figure. This was due largely to the great difference in the size of the organisms. A great mass of Clathrocystis was given no more weight in the result than a tiny Cyclotella. Each counted one, though the former sometimes contained a thousand times as much organic matter as the latter. In order to make the figures representing the total number of organisms bear some close relation to the actual character of the water as shown by the physical and chemical analyses, it was suggested that the standard unit already in use for the amorphous matter might be applied to the organisms as well. This "standard unit method" was adopted at the Boston Water Works, and has been used extensively elsewhere.

The unit system does not involve much extra labor in the counting. Many organisms are so constant in size that they may be counted individually and then reduced to standard units by multiplying by a constant factor. Filamentous forms of constant width may be measured in length and then reduced to units. Irregular masses and variable colonies may be estimated directly in units. In practice it has been found desirable to modify the unit somewhat in cases where organisms are especially thick or thin in order that the results may approximate a volumetric determination as nearly as possible.

It is not always that the unit system gives better results than the counting system. Sometimes it is advisable to state the results both in number of individuals and in standard units.

MICROSCOPICAL EXAMINATION.

Sample of Croton Water, New York.
Date of Collection, Aug. 25, 1897; *Date of Examination*, Aug. 25, 1897.
Concentration, 500 cc. to 10 cc.   *Factor*, 2.

Number of Square	1	2	3	4	5	6	7	8	9	10	Total	Number per c.c.	Standard Units per c.c.
DIATOMACEÆ:													
Asterionella				8		4					12	24	9
Cyclotella	1		3	1				2	5		16	32	3
Melosira	10	12	12	11	9	7	22	19	37	11	150	300	150
Navicula	2	5	8	4	2	7	6	4	7	6	51	102	102
Synedra	3	1	1	2	2		2	4	2	2	19	38	38
Tabellaria	6		7		10	8		5	6		42	84	84
Cymbella		1		2	1	2		4	3		13	26	26
Cocconeis	1		2		2		1			4	10	20	10
CHLOROPHYCEÆ:													
Closterium	1			1	1				1		4	8	40
Pediastrum		1	1			1	1			1	5	10	100
Protococcus	2		4	3		5	2	1	6	5	28	56	56
Scenedesmus	1	3	1	2		2		1	2	3	15	30	15
Staurastrum		1	1				1				3	6	18
Spirogyra					30						30	60	60
Pandorina	1	1				1	1				4	8	80
CYANOPHYCEÆ:													
Anabæna	80	90	160	130	240	70	110	150	100	110	1240	—	2480
Chroococcus			5				5				10	20	20
Clathrocystis					25						25	—	50
Cœlosphærium	40	30	75	40	25		90	75	35	50	470	—	940
Microcystis			10		10	10					30	—	60
FUNGI AND SCHIZOMYCETES:													
Crenothrix		2		1		1		2	2		8	16	16
Mold Hyphæ			2		2						4	8	8
Cladothrix						2					2	4	4
PROTOZOA:													
Dinobryon			10				8				18	36	18
Mallomonas					1			1			2	4	8
Peridinium	1	2			1		1				5	10	60
Synura			10			10					20	—	40
Trachelomonas	1		1	1		1		2			6	12	12
Cryptomonas		1			1						2	4	4
Codonella									1		1	2	16
Ceratium							1				1	2	20
ROTIFERA:													
Anuræa			1								1	2	40
Polyarthra					1						1	2	50
CRUSTACEA:													
Cyclops											Present		
OTHER ORGANISMS:													
Anguillula		1									1	2	10
TOTAL ORGANISMS												—	4647
AMORPHOUS MATTER	20	25	40	25	15	40	20	30	35	20	270	—	540
MISCELLANEOUS BODIES:													
Sponge spicules											3	6	6

METHODS OF MICROSCOPICAL EXAMINATION. 33

SCHEDULES OF CLASSIFICATIONS USED AT DIFFERENT TIMES AND IN DIFFERENT LABORATORIES.

Individual Counting System.				Standard Unit System.	
Mass. St. Bd. of Health. Parker, 1887.	Boston Water Works. Whipple, 1889.*	Mass. St. Bd. of Health. Calkins, 1890.	Conn. St. Bd. of Health. 1891.	Brooklyn Water Dept. Whipple, 1897.	Boston Water Works. Hollis, 1897.
Diatomaceæ	Diatomaceæ	Diatomaceæ	Diatomaceæ	Diatomaceæ	Diatomaceæ
Desmidieæ Palmellaceæ Zoosporeæ Zygnemaceæ Volvocinieæ	Desmidieæ Chlorophyceæ	Algæ	Desmidieæ Protococcoideæ Confervaceæ	Chlorophyceæ	Chlorophyceæ
Cyanophyceæ	Cyanophyceæ	Cyanophyceæ	Cyanophyceæ	Cyanophyceæ	Cyanophyceæ
Schizomycetes	Fungi	Fungi	Fungi	Fungi and Schizomycetes	Fungi
Protozoa	Rhizopoda Infusoria	Rhizopoda Infusoria	Rhizopoda Infusoria	Protozoa	Rhizopoda Infusoria
Rotifera	Rotifera	Vermes	Rotifera	Rotifera	Rotifera
Entomostraca	Crustacea	Crustacea	Crustacea	Crustacea	—
Spongiaria Nematoda Annelida	Miscellaneous	Miscellaneous (including Zoogloea)	Ova Spores	Other Organisms	Miscellaneous
—	Total Organisms	Total Organisms	—	Total Organisms	Total Organisms
	Amorphous Matter			Amorphous Matter	Amorphous Matter
				Miscellaneous Bodies	—

Records.—The results of analysis may be recorded on a blank similar to the one shown on page 32. The ten numbered vertical columns correspond to ten squares counted. The two right-hand columns give the results in " Number per c.c." and in " Number of Standard Units per c.c." Either

---

* The Standard Unit system has been used since Jan. 1, 1893.

or both of these columns may be used. The names of the common organisms are given in the left-hand column, and are grouped according to the system of classification described in Part II. The table on page 33 shows the schedules of classification used by different observers. It may be found useful in the comparison of different reports.

II. PLANKTON NET METHOD.

The plankton net * consists of a conical net of silk bolting-cloth (No. 20) suspended from an iron ring two feet in diameter (Fig. 7). The net has a length of three feet. At the lower end it terminates in a flat metal ring to which is attached the filtering-bucket. The latter consists of a metal frame covered on the sides with bolting-cloth, and having a slightly conical bottom. In the middle of the bottom there is an outlet-tube closed with a removable plug. The bucket is about 2½ inches in diameter. It is supported on three legs when detached from the net. The filtering-net of bolting-cloth is protected by a twine net which helps to bear the strain when the net is drawn through the water. Cords extend from the iron ring to the bucket in order to further relieve the filtering-net from strain. Above the filtering-net there is a truncated canvas cone that serves as a guard, preventing the entrance of mud when near the bottom and preventing the contents of the net from spilling over the edge. The smaller diameter of this guard is about 16 inches. It is this diameter that determines the volume of water filtered when the net is drawn through the water. The whole net is

---

* There are several modifications of Hensen's original net. The form used by Reighard in Lake St. Clair and here described may be considered as typical.

suspended by three cords attached to radiating iron arms fastened to the rope by which the apparatus is raised and lowered.

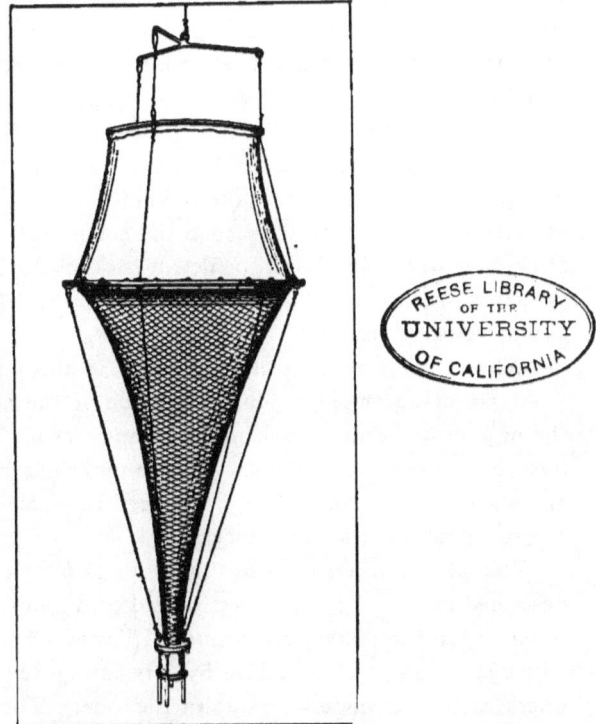

FIG. 7.—PLANKTON NET. (After Reighard.)

The net is operated as follows: It is lowered to the bottom or to the desired depth and then drawn to the surface, the velocity of its ascent being noted. On the way down it takes in no water except what is filtered through the gauze. On the way up it filters a column of water whose cross-section is that of the opening of the guard net and whose height is equal to the distance through which the net was drawn. This is the theoretical amount filtered. In practice the net does not strain the whole column of water through which it passes, as a portion of the water is forced aside. Therefore in order to obtain the volume of plankton in the column traversed it is necessary to multiply the observed result by a factor or coefficient. This net-coefficient varies for each net and for different velocities of ascent through the water. It also varies with the amount of clogging. With velocities of 2 to 3 ft. per second the coefficient is about $2\frac{1}{2}$. It is necessary to know the coefficient for each net at different velocities and to correct the results of each haul for the particular velocity used.

When the net reaches the surface it is allowed to drain. A stream of water played on the outside of the net detaches the organisms from the bolting-cloth and washes them down into the bucket. The bucket is then detached from the net and its collected material is transferred to a small bottle for transportation to the laboratory.

The plankton net used by Birge differs from the one just described in that it has a cover instead of a guard-net. The cover slides in a rectangular frame. It is moved by delicately adjusted weights set in action by a releasing device which is operated by messengers sent down the rope. The cover may be opened or closed at any depth at the will of the operator. This enables one to collect material from the lower strata without having it contaminated with that above it.

The amount of plankton collected may be determined by four methods: (1) by estimation of the volume; (2) by determination of the weight; (3) by chemical analysis; (4) by enumeration of the organisms.

The volume is obtained by allowing the material to stand in alcohol in a graduated cylinder for 24 hours. At the end of that time the plankton will have settled and the volume in cubic centimeters may be read from the scale. This gives the total volume in one catch. It is customary to express results in "number of cubic centimeters of plankton under one square meter of surface" or in "number of cubic centimeters of plankton in one cubic meter of water."

The approximate weight may be determined by drying on filter-paper and weighing. The results are usually expressed in grams of plankton under one square meter of surface or in one cubic meter of water.

The chemical analysis of the plankton usually consists of the determination of the percentage of organic material, ash, silica, etc.

The enumeration of the organisms is the most important part of the laboratory investigation. The material is evenly distributed in a definite amount of alcohol by shaking, and a portion is removed to a small trough or cell and placed under the microscope. The various organisms are then counted. Lines drawn on the bottom of the cell aid the observer in covering the entire area of the cell. As in the case of volume and weight, the results are generally expressed either in "number of organisms under one square meter of surface" or in "number of organisms per cubic meter of water." Both these methods are objectionable because so many figures are involved. They often extend to the millions and sometimes to the billions. It is preferable to express the smaller organisms, such as the algæ and protozoa, in "number per cubic centimeter," and the larger organisms, such as the crustacea, rotifera, etc., in "number per liter."

It is evident that the "plankton net method" involves

many sources of error. Neither the amount of water strained nor the completeness of the filtration can be definitely ascertained. The loss of the smaller organisms by leakage through the meshes of the silk is very great, and many of the delicate organisms are crushed upon the net. The methods of estimating the volume and weight of the plankton, moreover, are exceedingly inaccurate. The method of enumerating the organisms is much to be preferred. Except in the case of comparatively large organisms, such as the Rotifera, Crustacea, etc., the results of the net method cannot be depended upon within 50 per cent.

III. Plankton Pump.

The plankton pump is designed to collect the plankton from any particular depth in a lake. It consists of a sort of force-pump so arranged that a definite and measurable quantity of water is delivered at each stroke; an adjustable hose through which the water is drawn from the desired depth; and a filtering-bucket into which the water is pumped. The straining is effected by allowing the water to pass through a cylinder of fine wire gauze at the lower end of the filtering-bucket. The efficiency of the strainer is increased by covering the wire gauze with fine bolting-cloth.

This method has the advantage of measuring the quantity of water strained with greater accuracy than is possible in the net method, but the error from imperfect filtration is large.

IV. The Planktonokrit.

The planktonokrit is a modification of the centrifugal machine. The water to be examined is placed in two funnel-shaped receptacles attached to an upright shaft, with the necks of the funnels pointed outwards. The receptacles have a capacity of one liter each. The funnel portion is made of tinned copper; the stem is a glass tube that has a bore of $2\frac{1}{2}$ to 5 mm. The glasses are held in place by a cover, such as is employed in mounting a water-gauge. The shaft is driven by hand or belt through a series of geared wheels, so arranged that 50 revolutions of the crank, or pulley-wheel, produce 8000 revolutions of the upright shaft. By this rapid revolution of the sample the organisms are thrown outwards by centrifugal force and collect in the neck of the funnel, from which they may be removed for examination.

There are certain practical objections to the forms of apparatus now constructed. It is not only difficult but dangerous to use high speeds when large quantities of water are operated on. Field has been unable to use a speed greater than 3000 revolutions per minute. This speed maintained for four minutes, however, was sufficient to throw out all the organisms except the Cyanophyceæ. By reducing the amount of the samples and by perfecting the mechanical parts of the apparatus it seems probable that excellent results may be obtained by this method.

Comparison of the methods described above will show that the Sedgwick-Rafter method and the planktonokrit are designed for use in examining samples of water in the laboratory, while the plankton net and the plankton pump are intended for field work. The latter are most serviceable in concentrating the larger microscopic organisms such as the Rotifera and Crustacea. The Sedgwick-Rafter method is the most practical and efficient method for use in sanitary water analysis. It should not be relied upon completely, but should be supplemented by a direct microscopical examination of the original sample of water or by the use of the planktonokrit.

It is much to be desired that all results, obtained by whatever method, should be expressed in terms of the same unit, and it is hoped that the inconvenient methods of expressing results in "grams or cubic centimeters of plankton under one square meter of surface or in one cubic meter of water" will be abandoned by planktologists and the more exact system of counting the organisms substituted.

CHAPTER IV.

MICROSCOPIC ORGANISMS IN WATER FROM DIFFERENT SOURCES.

IN studying the distribution of microscopic organisms it will be convenient to consider the following classes of water-supply separately:

    RAIN-WATER.
    GROUND-WATER.
      Springs.
      Wells.
      Filter-galleries (Infiltration-galleries).
      Filter-basins (Infiltration-basins).
    SURFACE-WATER.
      Streams and Canals.
      Natural Lakes and Ponds.
      Artificial Reservoirs.

Rain-water.—Rain-water is perhaps the purest water found in nature, yet it sometimes contains micro-organisms. For the most part they are so minute that an examination by the Sedgwick-Rafter method fails to reveal them, but larger forms are sometimes observed.

The study of the organisms found in rain-water is really the study of the organisms found in the air. It is worthy of more attention than has been given to it. The presence of organisms, or their spores, in the air may be demonstrated by sterilizing some water rich in nitrogenous matter and exposing it to the air in the light. After a week or two it will contain numerous forms of microscopic organisms which must have settled into the liquid from the air or developed from spores floating in the air.

Rain-water collected in a sterilized jar and allowed to stand protected from the air often develops a considerable growth of algæ (usually some Protococcus form), showing that the rain has not only taken up the organisms or their spores, but has absorbed sufficient food material for their growth. Samples of rain-water sometimes contain a surprisingly large amount of nitrogenous matter, especially if collected in the vicinity of a large city and at the beginning of a storm.

It has been noticed frequently that vigorous growths of algæ have appeared in ponds or reservoirs immediately after a rain-storm, the growth occurring suddenly and simultaneously throughout the whole body of water. It is possible that these sudden growths may be caused by the dried spores of the algæ being lifted from the shores of the ponds and scattered through the air by the wind, and then washed into the water by the rain. This supposition is in harmony with the theory that in the case of certain algæ sporadic development occurs only after the desiccation of the spores.

Ground-water.—Ground-water is water that has filtered or percolated through the ground. It comes to the surface as springs or is collected in wells, filter-galleries, or filter-basins.

Ground-water collected directly from the soil before it has had an opportunity to stand in pipes or be exposed to the light is almost invariably free from microscopic organisms. Its passage through the soil filters them out. It usually contains an abundant supply of plant food, extracted from the organic and mineral matter of the soil and modified by bacterial action, and when the water reaches the light this food material is seized by the micro-organisms. One will recall the luxuriant aquatic vegetation at the mouth of some spring or in some watering-trough supplied with spring-water. Organisms are occasionally met with in ground-water supplies, but, with the exception of the Schizomycetes, the number of organisms depends upon the exposure of the water to the light and air; that is, it is only as a ground-water becomes a surface-water that the microscopic organisms develop.

The following table, compiled from the examinations of the Massachusetts State Board of Health, gives an idea of the organisms met with in ground-water supplies. Except in the case of springs, the figures represent the average of monthly observations extending over one or more years.

Spring-waters usually contain no microscopic organisms. Several exceptions are noted in the table,—one at Westport, where 455 Himantidium were present, and one at Millis, where the water contained 180 Chlamydomonas per cc. That these were accidental is shown by the fact that in 1893 five examinations of the Aqua Rex Spring showed an entire absence of organisms.

Well-waters also are ordinarily free from organisms, but in some cases Crenothrix grows abundantly in the tubes of driven wells. This is particularly true if the water is rich in iron and in organic matter. Wells driven in swamps are often thus affected. The tubular wells at Provincetown are an example. Crenothrix is sometimes found there as abundant as 20 000 per c.c. The water contains more than 0.125 parts of albuminoid ammonia per million, and the iron varies from 1.00 to 5.00 parts per million. Many similar cases might be cited. Leptothrix and Spirochæte forms are also observed in

MICROSCOPIC ORGANISMS IN GROUND-WATERS.
(NUMBER PER C.C.)

No.	Locality.	Time.	Diatomaceæ.	Chlorophyceæ.	Cyanophyceæ.	Fungi.*	Rhizopoda.†	Infusoria.†	Rotifera.	Total Organisms.	Zooglœa (Units).
	SPRING-WATERS.										
I	Spring in Westport	Apr. 21, 1894	455	0	3	0	0	0	1	459	0
II	Aqua Rex Spring, Millis	Aug. 27, 1894	1	180	0	0	0	0	0	181	0
III	Craig Spring, West Springfield	May 16, 1893	21	0	0	0	0	0	0	21	16
IV	Spring in Ipswich	July 27, 1892	12	0	0	0	0	0	0	12	0
V	Spring in Pepperell	Nov. 16, 1894	2	1	0	2	0	0	0	4	0
VI	Massasoit Spring, West Springfield	May 16, 1893	0	0	0	2	0	0	0	2	0
VII	Spring in Ware	July 17, 1893	0	0	0	0	0	1	0	1	0
VIII	Spring in Medfield	Aug. 31, 1894	0	0	0	0	0	0	0	0	0
IX	Spring in Pittsfield	Aug. 27, 1894	0	0	0	0	0	0	0	0	0
X	Cold Spring, Plymouth	July–Dec. 1894	0	0	0	0	0	0	0	0	0
	WELL-WATERS.										
I	Tubular Well, Provincetown	1894	0	0	0	3130	0	0	0	3130	50
II	Tubular Wells, Revere	1894	1	0	0	281	0	0	0	282	—
III	Large Collecting Well, Marblehead	1894	0	0	0	173	0	0	0	173	8
IV	Tubular Wells, Hyde Park	1893–4	2	0	1	68	0	0	0	70	18
V	Tubular Wells, Malden	1891–3	5	2	0	1	0	pr.	0	8	7
VI	Tubular Wells, Lowell	1893	0	0	0	0	pr.	1	0	2	—
VII	Tubular Wells, Melrose	1894	0	0	0	1	0	0	0	1	—
VIII	Tubular Wells, Bradford	1893	0	0	0	0	0	pr.	0	pr.	547
IX	Well, Needham	1894	0	0	0	0	0	0	1	1	0
X	Well at Fitzwilliam, N. H.	1893	0	0	0	0	0	0	0	0	0
	FILTER-GALLERIES. (Infiltration-galleries.)										
I	Filter-gallery at Reading	1891–4	3	0	0	3506	0	2	0	3511	726
II	Filter-gallery at Wayland	1891	15	4	1	1706	0	3	0	1729	71
III	Filter-gallery at Whitman	1891	pr.	0	0	137	0	0	0	138	41
IV	Filter-gallery at Watertown	1892	1	0	0	217	0	0	0	217	72
V	Filter-gallery at Framingham	1891	0	0	0	137	0	2	0	138	41
VI	Filter-gallery at Braintree	1894	2	0	0	34	0	0	0	36	94
VII	Filter-gallery at Woburn	1891	0	2	4	0	0	0	1	2	1
VIII	Filter-basin at Taunton	1891–4	86	1	0	24	0	48	0	165	14
IX	Filter-basin at Newton	1892–4	2	0	0	15	0	pr.	0	18	13
X	Filter-basin at Waltham	1892	17	0	0	0	0	pr.	0	17	4

* Including the Schizomycetes.   † Protozoa.

well-waters rich in iron. Crenothrix grows in tufts or in felt-like layers on the inner walls of the tubes. By the deposition of iron oxide in its gelatinous sheath it clogs up the tubes and strainers with iron-rust.

Filter-galleries, or infiltration-galleries, are practically elongated wells located near some stream or pond. They are similar to wells in regard to the presence of micro-organisms. Few organisms other than Crenothrix are found.

Filter-basins, or infiltration basins, are filter-galleries open to the light. The water in them is sometimes affected with algæ-growths. The filter-basin at Taunton, Mass., for example, has given trouble from this cause. In October 1894 there were more than 1000 Asterionella per c.c. present, and they were followed by a vigorous growth of Dinobryon. Filter-basins are practically open reservoirs for the storage of ground-water—a subject to be treated in another chapter.

Surface-water.—The term "surface-water" includes all collections of water upon the surface of the earth, i.e., lakes, ponds, rivers, pools, ditches, etc.

The following table shows that surface-waters contain many more microscopic organisms than ground-waters, and that standing water contains more organisms than running water.

Samples from rivers, unless collected near the shore, seldom contain many organisms, and water-supplies drawn from rivers and subjected to limited storage are not often troubled with animal or vegetable growths. This may be true even where the banks of the stream are covered with aquatic vegetation. The organisms found in streams are largely sedentary forms. Their food-supply is brought to them by the water continually passing. In quiet waters there are found free-swimming forms that must go in search of their

MICROSCOPIC ORGANISMS IN SURFACE WATERS.

(NUMBER PER C.C.)

No.	Locality.	Time.	Diato-maceæ.	Chloro-phyceæ	Cyano-phyceæ	Fungi.*	Rhizo-poda.†	Infu-soria.†	Ro-tifera.	Total Organisms.	Zooglœa (Units).
	RIVERS.										
I	Stony Brook, Influent to Basin 3	1891 2	77	23	43	38	1	9	0	191	97
II	Mill River at Taunton	July–Sept. 1893	3	25	1	165	1	4	pr.	199	676
III	Merrimac River at Lawrence	1891–4	66	21	2	13	pr.	4	pr.	106	156
IV	Ipswich River	1892	12	1	0	87	0	5	0	105	31
V	Blackstone River at Uxbridge	1892	17	6	0	3	0	5	pr.	100	394
VI	Sudbury River, Influent to Basin 2	1891 2	45	16	0	32	pr.	74	pr.	98	128
VII	Cold Spring Brook, Influent to Basin 4	1891	54	pr.	2	12	0	3	0	77	39
VIII	Nashua River, North Branch	1893	13	4	2	1	0	6	0	67	810
IX	Taunton River	1893–3	17	1	2	42	0	2	0	35	58
X	Lynde Brook, Worcester	1891	17	4	3	2	0	1	0	17	68
	NATURAL PONDS.										
I	Mystic Lake	1891–4	1017	199	pr.	18	pr.	172	pr.	2306	128
II	Jamaica Pond	Jan.–Aug. 1891	1110	103	137	1	1	12	1	1305	174
III	Horn Pond, Woburn	1891–4	911	302	218	1	1	167	2	1602	65
IV	Fresh Pond, Cambridge	1891–4	967	95	83	9	1	4	pr.	1159	17
V	Wenham Lake, Salem	1891 4	897	38	32	0	pr.	32	pr.	999	52
VI	Buckmaster Pond, Norwood	1891–4	134	83	9	2	2	665	pr.	944	30
VII	Lake Cochituate	1891–4	579	33	58	6	2	15	pr.	693	66
VIII	Spot Pond, Malden	1891–4	171	85	19	1	1	19	pr.	296	93
IX	Lake Williams, Marlboro	1891	170	17	66	1	0	14	0	268	67
X	Gates Pond, Hudson	1891–4	110	37	27	1	1	66	pr.	242	38
	ARTIFICIAL RESERVOIRS.										
I	Haynes Reservoir, Leominster	1891	3193	0	0	1	0	19	1	3214	155
II	Walden Pond, Lynn	1891–4	254	238	604	8	pr.	397	pr.	1502	64
III	North Reservoir, Winchester	1891–4	1337	35	72	5	1	149	pr.	1596	71
IV	Ludlow Reservoir, Springfield	1891–4	504	260	96	2	1	96	2	964	103
V	Scott Reservoir, Fitchburg	1892	693	146	10	1	4	92	1	947	42
VI	Holden Reservoir, Worcester	1891–4	646	24	6	1	pr.	29	pr.	707	76
VII	Basin 3, Boston	1891–4	270	55	23	5	1	12	1	362	122
VIII	Basin 2, Boston	1891–4	99	32	47	1	pr.	4	pr.	187	76
IX	Basin 4, Boston	1891–4	80	31	3	1	0	5	0	120	125
X	Basin 6, Boston	1894	55	5	0	0	1	31	2	94	43

* Including the Schizomycetes.   † Protozoa.

food. It is difficult to draw a sharp line between these two classes of organisms. Some are free-swimming at will or during a part of their life-history, and some free-swimming organisms are always found associated with sedentary forms. On most rivers there are some quiet pools where free-swimming forms may develop. In a sample of river-water, then, one is likely to find sedentary forms which have become detached, organisms which have developed in the quiet places or in tributary ponds, and spores or intermediate forms in the life-history of sedentary organisms. In streams draining large ponds or lakes the water naturally has the character of the pond- or lake-water, and organisms may be abundant.

The number of microscopic organisms found in rivers is subject to great fluctuations. If the water is rich in food-material, littoral growths often develop with rapidity, while a heavy rain that increases the current of the water and the amount of scouring material that it carries may suddenly wash away the entire growth. With such conditions the number of organisms collected in a sample may be above the normal. At other times a rain may diminish the number of organisms in a sample by dilution. But the fluctuations are due chiefly to changes that take place in the growths in tributary ponds or swamps, and to the fact that rains may cause these ponds to overflow.

The table shows that the Diatomaceæ are the organisms found most constantly in rivers. Navicula, Cocconema, Gomphonema and other attached forms are common, but their numbers are small compared to those found in standing water. Some of the Chlorophyceæ, particularly Conferva, Spirogyra, Draparnaldia and other filamentous forms, are often observed. The Cyanophyceæ, except the Oscillarieæ, seldom occur. Stony Brook, in the table, represents a stream affected by tributary ponds where Cyanophyceæ abound. Crenothrix is quite often found in river-water. Anthophysa is often mistaken for it, and this may account in part for the high figures in the table. Animal forms are not common in rivers unless the water is polluted. Rotifera and Crustacea are seldom seen, but Protozoa are sometimes observed.

In the slow-running water of canals and ditches organisms sometimes develop in large numbers, but the conditions are not often such as to cause trouble in public water-supplies. The following instance, however, is worth noting:

On Sunday, July 12, 1896, it was observed by some of the residents living in the western part of the city of Lynn, Mass., that the water drawn from the service-taps had a green color. A glass of it showed a heavy green sediment when allowed to stand even for a few minutes. On the following day it became worse, and when the water was used for washing in the laundry it was found to leave green stains on the clothes. These acted like grass-stains. Investigation showed that the stains were caused by Raphidomonas, and that these organisms were abundant in the city water. Examination of the four storage-reservoirs showed that they were not present there in sufficient numbers to account for the trouble. The water from one of the supply-reservoirs, Walden Pond, reaches the pumping-station by means of an open canal, tunnel, and pipe-line. It was in this open canal that the Raphidomonas were found. The sides of the canal were thickly covered with filamentous algæ, chiefly Cladophora. The water in the canal had a dark green color. When a bottle of it was held to the light it was almost opaque and was seen to be densely crowded with moving green organisms. As many as 2000 per c.c. were present. Evidently the organisms had developed among the algæ in the canal and had gradually scattered themselves out into the water from Walden Pond as it passed through the canal on its way to the city. The trouble was remedied by emptying the canal through the wasteways and cleaning the slopes to prevent later development.

This is the only case on record where Raphidomonas has caused trouble, though the organism is often found in surface-water supplies.

Quiescent Waters. — All quiescent surface-waters are liable to contain microscopic organisms in considerable numbers. The water that is entirely free from them is very rare. It is scarcely possible to collect a sample of stagnant water at any season of the year without obtaining one or more forms of microscopic life. The extent and character of the growths vary greatly in different ponds and at different seasons.

As it is in ponds and lakes and reservoirs that the microscopic organisms cause the most trouble, it is these bodies of water that chiefly interest us. Before considering the organisms in this class of water-supplies it is important to know something about the physical conditions of water in ponds and lakes. These are discussed in the following chapter. In passing, one should observe from the table that all classes of organisms, except perhaps the Schizophyceæ, are much more abundant in natural ponds and in reservoirs than in rivers.

CHAPTER V.

LIMNOLOGY.

LIMNOLOGY is that branch of science that treats of lakes and ponds,—their geology, their geography, their physics, their chemistry, their biology, and the relations of these to each other. This subject has taken shape only within the past ten years, but already many valuable publications have appeared.

In this chapter it is possible to consider only such limnological studies as are closely related to the microscopic organisms. The most important of these are: the temperature of the water, the amount of light received and transmitted by the water, and the food material of the organisms found in the water. The location of lakes, their shape, size, and depth, the source of their supply, the character of the watershed, the meteorology of the region,—all have their effect upon the organisms living in the water, but they can be considered only incidentally.

Physical Properties of Water.—The density of water varies with its pressure, with its temperature, and with the substances dissolved in it.

Grassi gives the coefficient of compressibility of pure water as .0000503 per atmosphere at 0° C., and .0000456 at 25° C. Therefore if the density at the surface of a lake is unity, at a depth of 339 ft. (10 atmospheres) it will be 1.0005;

at 678 ft. (20 atmospheres), 1.001; and at 1017 ft. (30 atmospheres), 1.0015.

Water attains its maximum density at about 4° C. or 39.2° F. Assuming its density at 4° C. to be unity, its density at other temperatures is given in the following table.

DENSITY OF WATER AT DIFFERENT TEMPERATURES.

Temperature.		Density.	Temperature.		Density.
Centigrade.	Fahrenheit.		Centigrade.	Fahrenheit.	
0°	32.0°	.99987	18.3°	65.0°	.99859
1.6	35.0	.99996	21.1	70.0	.99802
4.0	39.2	1.00000	23.8	75.0	.99739
4.4	40.0	.99999	26.6	80.0	.99669
7.2	45.0	.99992	29.4	85.0	.99592
10.0	50.0	.99975	32.2	90.0	.99510
12.7	55.0	.99946	35.0	95.0	.99418
15.5	60.0	.99907	37.7	100.0	.99318

Water freezes at 0° C., or 32.0° F. Ice is lighter than water. It readily floats in water at 0° C.

Water has a very high specific heat. It is a poor thermal conductor. Prof. W. H. Weber* gives its coefficient of conductivity as 0.0745.

Water is extremely mobile. This property renders it subject to displacement by mechanical agencies, such as wind and currents (mechanical convection), and permits it to become stratified according to the density of its particles. The mobility of water varies somewhat with its temperature, being greater as the temperature is higher.

When water is stratified with the warmer layers above the colder, the stratification is said to be "direct." This occurs when the temperatures are above that of maximum density. When water is stratified with the colder layers above the warmer the stratification is said to be " inverse." This

---

* Vierteljahreschrift der Zürich Nat. Ges., XXIV. 252, 1879.

occurs when the temperatures are below that of maximum density. With the temperatures above 39.2° it sometimes happens in a deep lake that a colder layer of water is found above a warmer layer. This is a paradox theoretically possible, because the density of the water at any point in a lake depends upon its depth as well as its temperature. Thus, water at 45° F. has a density of .99992. If this water were at a depth of 1017 ft., where the pressure is 30 atmospheres its density would be .99992 + .0015 = 1.00142, i.e., more than that of water at 39.2° F. at the surface. In nature, however, such a condition of temperatures seldom exists for a long period, and practically represents a state of unstable equilibrium. A thermal paradox may be caused also by differences in the density of different strata due to substances in solution.

Water has a slight power of diathermancy, i.e., it permits the penetration of radiant heat to a slight degree. Forel experimented on the diathermancy of water by comparing the readings of thermometers with blackened and with ordinary bulbs at a depth of 1 metre. He obtained the following results:

Date.	Time of Exposure.	Temperature of Water. (Fahrenheit.)	Excess of Temperature of Black Bulb Thermometer, in Fahr. Deg.
Mar. 27, 1871	10 hours	44.4°	10.8°
July 25, 1873	17 "	72.0	14.0
" 26, 1873	15 "	74.3	15.3
Aug. 1, 1873	12 "	75.2	7.6

THE TEMPERATURE OF LAKES AND PONDS.

Methods of Observation.—The observation of the temperature of the water at the surface of a lake is a comparatively easy matter, but it requires an accurate thermometer and a careful observer. Where the water is smooth the thermometer-bulb may be immersed just beneath the surface in an inclined position and the reading taken before removing it from the water. In taking the reading one must be careful to avoid parallax by holding the thermometer exactly at right angles to the line of sight. When the water is too rough for reading directly some of the surface-water may be dipped up and the temperature of that ascertained. Thermometers with bulb immersed in a cup are prepared for this purpose. Direct observations are much to be preferred.

The observation of the temperature of the water at depths below the surface is more difficult.

The simplest method of obtaining results that are in any way accurate is to enclose a weighted thermometer in a stoppered empty bottle and to lower this to the proper depth and fill it by drawing out the stopper. After allowing a sufficient time for the apparatus and thermometer to acquire the exact temperature of the water the bottle is drawn to the surface and the reading taken before the thermometer is removed from the bottle. If the bottle is of sufficient size, if it is allowed to remain down long enough, if it is drawn rapidly to the surface and the reading taken at once, the error ought not to exceed one degree Fahrenheit. This method is impracticable for lakes much deeper than 50 ft., and beyond that depth some form of deep-sea thermometer is necessary. Several forms of maximum and minimum thermometers and of self-setting thermometers have been devised. The Negretti and Zambra thermometers have been used extensively for obtaining the temperature of very deep water. Several forms of electrical thermometers have been suggested, but the thermophone* is the only one that has proved of practical value.

---

*Invented and patented by H. E. Warren and G. C. Whipple.

54   THE MICROSCOPY OF DRINKING-WATER.

The thermophone (see Fig. 8) is an electrical thermometer of the resistance type. It is based upon the principle that the resistance of an electrical conductor changes with its temperature and that the rate of change is different for differ-

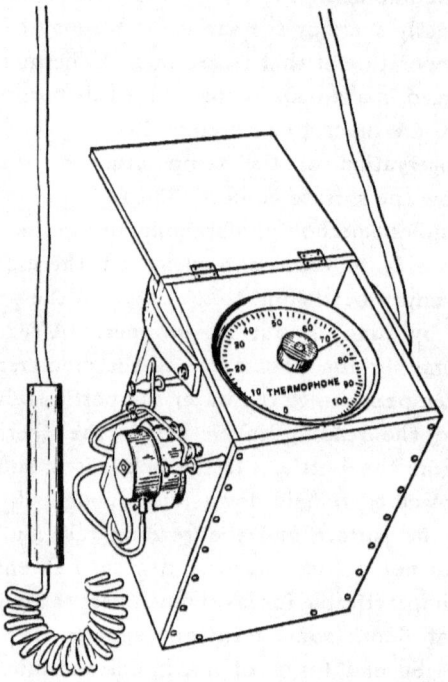

FIG. 8.—THERMOPHONE, PORTABLE FORM. (After Warren and Whipple.)

ent metals. Two resistance-coils of metals that have different electrical temperature-coefficients, as copper and German silver, are put into adjacent arms of a Wheatstone bridge and located at the place where the temperature is desired, the two coils being joined together at one end. The other extremities of the coils are connected by leading wires to the terminals of a slide-wire which forms a part of the indicator. A third leading wire extends from the junction of the two coils to a movable contact on the slide-wire, having in its circuit a telephone and a current-interrupter,—the latter operated by an independent battery connection. The telephone and interrupter serve as a galvanometer to detect the presence of a current. The slide-wire is wound around the periphery of a mahogany disc, above which there is another disc carrying a dial graduated in degrees of temperature. The movable contact which bears on the slide-wire is attached to a radial arm placed directly under the dial-hand, the two being moved together by turning an ebonite knob in the centre of the dial. This indicator is enclosed in a brass case in a box that also contains the batteries. The sensitive coils are enclosed in a brass tube of small diameter which is filled with oil, hermetically sealed, and coiled into a helix. Connections with the leading wires are made in an enlargement at one end. The leading wires are three in number and are made to form a triple cable. The temperature of the leading wires does not affect the reading of the instrument because two of them are of low resistance and are on opposite sides of the Wheatstone bridge. They neutralize each other. The third leading wire is connected with the galvanometer and does not come into the question. The readings of the instrument are independent of pressure.

The operation of taking a reading is as follows: The coil is lowered to the depth where the temperature is desired, the three leading wires are connected to the proper binding-posts of the indicator-box, the current from the battery is turned on, the telephone is held to the ear, and the index moved back and forth over the dial. A buzzing sound will be heard in the telephone, increasing or diminishing as the index is made to approach or recede from a certain section of the dial. A point may be found at which there is perfect silence in the telephone, and at this point the hand indicates the temperature of the distant coil. With thermophones adjusted for atmospheric range, i.e., from — 15° to 115° F., readings correct to 0.1° F. may be made. With a smaller range greater precision may be obtained.

Because of its accuracy, because of the ease with which the coil may be placed at any depth from the surface to the bottom of a lake, because of its extreme sensitiveness and rapidity of setting (one minute is sufficient), and because of its portability, the thermophone is better adapted than any other instrument for taking series of temperature observations in lakes at various depths. It has been used for that purpose at depths as great as 400 ft., and it was used by Prof. A. E. Burton in Greenland at much greater depths for obtaining temperatures in the crevasses of glaciers.

Results of Observations. — The temperature changes that take place in a body of water may be illustrated by a

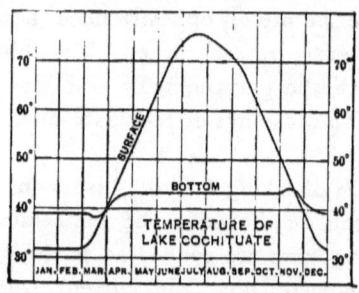

FIG. 9.

diagram that shows the temperatures at the surface and bottom of Lake Cochituate. The curves of Fig. 9 are based on a seven-years series of weekly observations, but some irregularities have been omitted for the sake of simplicity.

If one traces the line of surface temperatures, he will observe that during the winter the water immediately under the ice stands substantially at 32° F., though the ice itself often becomes much lower than 32° at its upper surface. As soon as the ice breaks up in the spring the temperature of the water begins to rise. This increase continues with some fluctuations until about the first of August. Cooling then begins and continues regularly through the autumn until the lake freezes in December. If this curve of surface temperature were compared with the mean temperature of the atmosphere for the same period a striking agreement would be noticed, and it would be seen that the water temperature is the higher of the two. When the surface is frozen there is no comparison between the air and water temperatures. During the spring and early summer, when the water is warming, the water is but slightly warmer than the air,* but during the late summer and autumn it is about 5° warmer. The surface temperature of the water fluctuates with the air temperature during the course of the day as well as on different days. The maximum is usually obtained between 2 and 4 P.M. and the minimum between 5 and 7 A.M. The daily range is seldom greater than 5°, though it may be much more. At the latitude of Boston the maximum surface temperature of the water of lakes during the summer is seldom above 80°.†

---

* It must be understood that it is the mean temperature of the air during 24 hours that is referred to, and not the maximum temperature during the daytime.

† A surface temperature of 92° was observed by the author at Chestnut Hill Reservoir on Aug. 12. 1896, at 3 P.M., after a week of excessively hot weather, during which the maximum daily temperature remained above

In small shallow ponds the surface temperature follows the atmospheric temperature much more closely than in large deep lakes where the water circulates to considerable depths. In the latter the surface temperature is often below that of the mean atmospheric temperature during the early part of the summer, and occasionally during the entire summer.

Lake Cochituate is 60 ft. deep. The temperature at the bottom during the winter, when the surface is frozen, is not far from that of maximum density (39.2° F.). The heaviest water is at the bottom; the lightest is at the top; and the intermediate layers are arranged in the order of their density. With these conditions the water is in comparatively stable equilibrium. It is *inversely* stratified. It is the period of "winter stagnation."

As soon as the ice has broken up in the spring the surface-water begins to grow warmer. Until it reaches the temperature of maximum density it grows more dense as it grows warmer, and as it becomes denser it tends to sink. Thus until the water throughout the vertical has acquired the temperature of maximum density there are conditions of unstable equilibrium caused by diurnal fluctuations of temperature that result in the thorough mixing of all the water in the lake. These conditions, together with the mechanical effect of the wind, usually cause a slight temporary lowering of the bottom temperature at this season. Finally the tem-

---

90°, while the humidity varied from 62% to 95%. At the time of the observation the air temperature was 95° and the humidity 70%. The temperatures of the water below the surface were as follows:

Surface	92.0°	10 ft	76.2°
1 ft	91.5	15 "	74.0
2 "	89.2	20 "	65.7
3 "	85.6	25 "	54.5
4 "	80.2	27 "	53.1
5 "	79.0		

perature throughout the vertical becomes practically uniform, and vertical currents are easily produced by slight changes in the temperature of the water at the surface and by the mechanical effect of the wind.

This is the period of " spring circulation " or the " spring overturning." It lasts several weeks, but varies in duration in different years. As the season advances the surface-water becomes warmer than that at the bottom, and finally the difference becomes so great that the diurnal fluctuation of surface temperature and the effect of the wind are no longer able to keep up the circulation. Consequently the bottom temperature ceases to rise, the water becomes " directly stratified," and the lake enters upon the period of " summer stagnation." During this period, which extends from April to November, the bottom temperature remains almost constant, and the water below a depth of about 25 ft. remains stagnant. In the autumn the surface cools and the water becomes stirred up to greater and greater depths, until finally the " great overturning " takes place and all the water is in circulation. At this time there is a slight increase in the bottom temperature that corresponds to the temporary lowering of the temperature in the spring. Then follows the period of " autumnal circulation," during which the surface and bottom strata have substantially the same temperature. In December the lake freezes and " winter stagnation " begins.

The use of the thermophone for obtaining series of temperatures at frequent intervals in the vertical has enabled the author to study the temperature changes in more detail, and to see how they are affected by the geography of the lake and the meteorology of the region.

In a frozen lake the water in contact with the under surface of the ice stands always at 32° F. The temperature at the bottom varies with the depth and with the meteorological conditions at the time of freezing. In most lakes, and particularly in deep lakes, it stands at the point of maximum density; in shallow lakes it may be lower than that; under abnormal conditions, as referred to on page 52, it may be slightly higher. During the period of winter stagnation the bottom temperature sometimes rises very slightly on account of direct heating by the sun's rays. This is because of the diathermancy of the water. The temperatures of the water between the surface and the bottom are illustrated by Fig. 10.

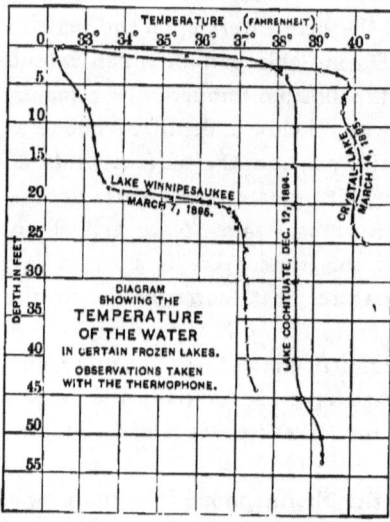

FIG. 10. (After FitzGerald.)

The cold water is usually confined to a thin layer—seldom more than 5 or 10 ft. thick—under the ice, and below that layer the temperature changes but little to the bottom. This is shown by the Lake Cochituate curve. This and the (abnormal) change in the curve at the bottom may be explained as follows: During the period of autumnal circulation the temperature is uniform throughout the vertical. As the weather gets colder the temperature throughout the vertical drops. Until the temperature has reached the point of maximum density the circulation of the water through the vertical takes place by thermal convection. Below that temperature it takes place chiefly by wind action. If the wind is not sufficiently strong to induce complete circulation the bottom temperature ceases to fall at 39.2°. Thus the bottom temperature at Lake Cochituate in December, 1894, was left at that point. Later the wind stirred the water to a depth of 45 ft., and above that depth the temperature became uniform at about 38.5°.

Freezing usually occurs on a cool, still night. The surface-water cools and freezes before the wind has had a chance to mix it with the warmer water below. The suddenness with which a lake freezes and the intensity of the wind at the time determine the depth of the layer of cold water, and the temperature of the air and the intensity of the wind previous to the time of freezing determine the temperature of the water at the bottom. The Lake Winnipesaukee curve (Fig. 10) represents the effect of a current flowing between two islands. A layer of cold water about 18 ft. thick was flowing over a quiet body of warmer water. The dividing line, at a depth of about 20 ft., was very sharply defined. The Crystal Lake curve (Fig. 10) shows abnormal conditions produced by springs at the bottom of the lake.

During the summer the temperature of the water is similarly affected by meteorological conditions. After the ice has broken up the temperature of the water at all depths rises. Above 39.2° circulation takes place chiefly by the action of the wind. If there were no wind, or if the wind were not sufficient, the temperature at the bottom would not rise above 39.2°. In very deep lakes this happens, but in most lakes the wind causes it to rise somewhat above that point. It continues to rise as long as the difference in density between the water at the surface and at the bottom does not become too great for the wind to keep up the circulation. In Lake Cochituate this difference of density is produced by a difference of about 5° in temperature. When stagnation has once begun the temperature at the bottom changes very little during the summer. It sometimes rises slightly on account of direct heating, as it does in the winter. If warm weather occurs early and suddenly in the spring the required difference of temperature between the upper and lower layers is soon obtained, and consequently the temperature at the bottom through the summer remains low. But if the season advances slowly the bottom temperature will become fixed at a higher point. In Lake Cochituate the bottom temperature varies in different years from 42° to 45°.

The temperatures of the water between the surface and bottom during the summer may be illustrated by the two typical curves in Fig. 11. Previous to May 13, 1895, the season had progressed gradually. On that day the atmospheric temperature rose to 90° and there was little wind. These conditions produced a uniform curve. Then followed several days of cold, windy weather. The surface temperature fell and the water became stirred to a depth of about 17 ft. Below 20 ft., however, there was little change. These conditions usually continue through the summer, the upper layers becoming warmed and stratified, or cooled and mixed, the lower layers remaining stagnant. Between these upper and lower layers there is a thin layer where the temperature changes rapidly,—sometimes 10° in one vertical foot. This region is sometimes called the *thermocline*. Its position and temperature gradient vary according to the depth of the lake, the intensity of the wind, and the temperature of the water above and below. The upper boundary of the thermocline is sometimes very sharp—particularly in the autumn; the lower boundary is less distinct. In the fall the position

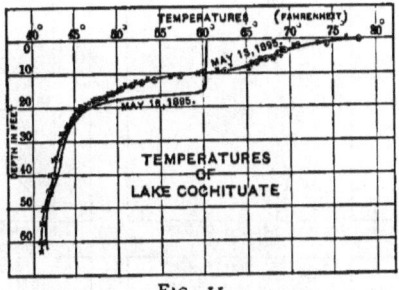

Fig. 11.

of the thermocline drops towards the bottom as circulation extends to greater and greater depths.

These seasonal changes of temperature are modified somewhat in very deep and in very shallow lakes and in lakes situated in extremely hot or cold climates, and these modifications may be used as a basis for classification.

Classification of Lakes According to Temperature.—Lakes may be divided into three *types*, according to their surface temperatures, and into three *orders*, according to their bottom temperatures. The resulting nine classes are shown in Fig. 12. On these diagrams the boundaries of the shaded areas represent the limits of the temperature fluctuations at different depths. The horizontal scale represents temperatures in Fahrenheit degrees increasing towards the right, and the vertical scale represents depth. The three types of lakes are designated as *polar*, *temperate*, and *tropical*. In lakes of the polar type the surface temperature is never above that of maximum density; in lakes of the tropical type it is never below that point; in lakes of the temperate type it is

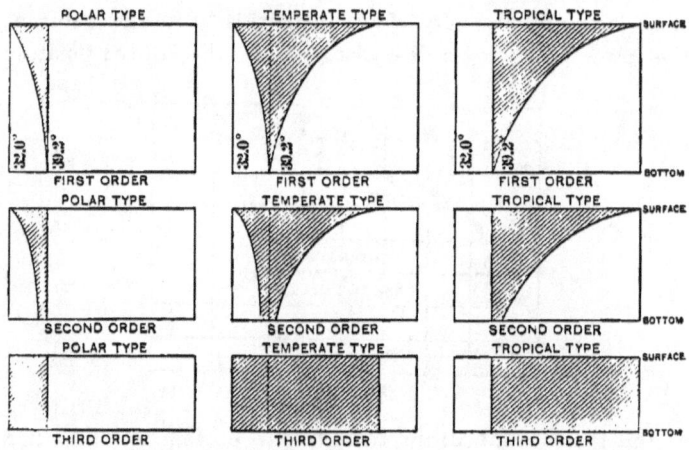

FIG. 12.—CLASSIFICATION OF LAKES ACCORDING TO TEMPERATURE.

sometimes below and sometimes above it. This division into types corresponds somewhat closely with geographical location.

The three orders of lakes may be defined as follows: lakes of the first order have bottom temperatures which are practically constant at or very near the point of maximum density; lakes of the second order have bottom temperatures which undergo annual fluctuations, but which are never very far from the point of maximum density; lakes of the third order have bottom temperatures which are seldom very far from the surface temperatures. The division into orders corresponds in a general way to the character of the lakes; i.e., their size, contour, depth, surrounding topography, etc.

The temperature changes which take place in the nine classes of lakes according to this system of classification are exhibited in another manner in Fig. 13. These diagrams show by curves the surface and bottom temperatures for each season of the year, the dates being plotted as abscissæ, and the temperatures as ordinates. The shaded areas show the difference between the surface and bottom temperatures,

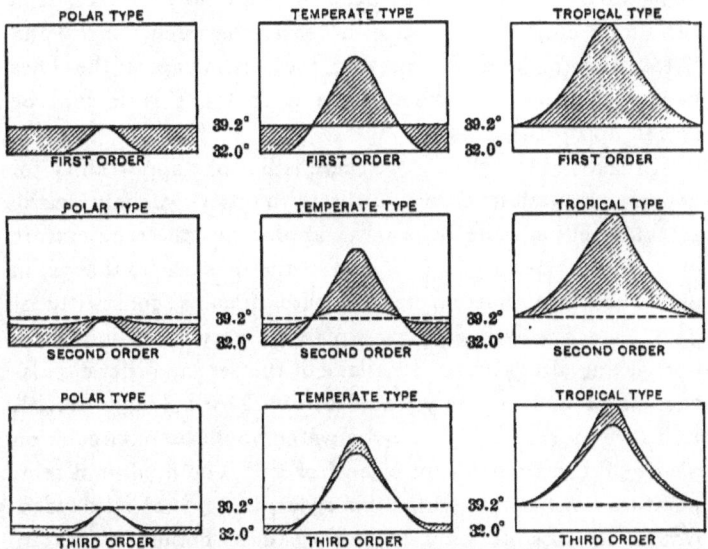

FIG. 13.—CLASSIFICATION OF LAKES ACCORDING TO TEMPERATURE.

the wider the shaded area the greater being the difference.

A study of these diagrams brings out some interesting facts concerning the phenomena of circulation and stagnation. In Fig. 12 it will be seen that the circulation periods occur when the curve showing the temperatures at various depths becomes a vertical line; that is, when the water all has the same temperature. The stagnation periods are shown by the line being curved, the top to the right when the warmer layers are above the colder, and to the left when the colder layers are above the warmer. In Fig. 13 the circulation periods are indicated by the surface and bottom temperature curves coinciding, and the stagnation periods by these lines being apart. The distance between the lines indicates, to a certain extent, the difference in density between the top and bottom layers, and we see that the farther apart the lines become the less likelihood there is that the water will be stirred up by the wind.

In lakes of the polar type there is but one opportunity for vertical circulation (except in the third order); namely, in the summer season, when the water approaches the temperature of maximum density. In a lake of the first order, that is, in one where the bottom temperature remains constantly at $39.2°$, the circulation period would be very short indeed, if not lacking altogether. In a lake of the second order circulation might and probably would continue for a longer period. In a lake of the third order the water would be in circulation nearly all the time except when frozen. The minimum temperature limit indicated for this order, i.e., $32°$ at all depths, would be possible only in very shallow bodies of water, and would simply indicate that all the water was frozen The temperature of the ice would probably be below $32°$ at the surface. It is probable that very few polar lakes exist.

In lakes of the tropical type there is likewise but one period of circulation each year (except in the third order). This would occur not in summer, but in winter. In the first order this circulation period would be brief or entirely wanting; in the second it would be of longer duration; in the third order the water would be liable to be in circulation the greater part of the year. Tropical lakes are quite numerous, but observations are lacking to place them in their proper order.

Most of the lakes of the United States belong to the temperate type. In this type there are two periods of circulation and two periods of stagnation (except in the third order), as we have seen illustrated in the case of Lake Cochituate. In lakes of the first order the circulation periods would be very short or entirely wanting; in the second order the circulation periods would be of longer duration; in the third order the water would be in circulation throughout the year when the surface was not frozen. The above facts may be recapitulated in tabular form as follows:

CIRCULATION PERIODS.

	Polar Type.	Temperate Type.	Tropical Type.
First Order.	One circulation period possible, in summer, but generally none.	Two circulation periods possible, in spring and fall, but generally none.	One circulation period possible, in winter, but generally none.
Second Order.	One circulation period, in summer.	Two circulation periods, in spring and autumn.	One circulation period, in winter.
Third Order.	Circulation at all seasons, except when surface is frozen.	Circulation at all seasons, except when surface is frozen.	Circulation at all seasons.

Speaking in very general terms, one may say that lakes of the first order have no circulation, lakes of the third order have no stagnation (except in winter); and lakes of the second order have both circulation and stagnation.

In view of the comparatively few series of observations of the temperature of our lakes, the author refrains from making any classification of the lakes of the United States, but the results thus far obtained seem to indicate that the first order will include only those lakes more than about two hundred feet in depth, such, for instance, as the Great Lakes, Lake Champlain, etc.; the second order will include those with depths less than about two hundred feet, but greater than about twenty-five feet; and the third order will include those with depths less than twenty-five feet. These boundaries are only approximate, and it should be remembered that depth is not the only factor which influences the bottom temperature.

Stagnation is sometimes observed in small artificial reservoirs even when the depth is less than twenty feet. It is usually of short duration.

Transmission of Light by Water.

The amount of light received by the micro-organisms in a lake depends upon the intensity of the light at the surface of the water and upon the extent to which the light is transmitted by the water. The transmission of light by water varies chiefly with the amount of dissolved and suspended matter that it contains. The former affects its coefficient of absorption; the latter acts as a screen to shut out the light. In studying the penetration of light into a body of water it is necessary to take account of its color and its turbidity.

Color of Water.—Some surface-waters are colorless, but in most ponds and lakes the water has a more or less pronounced brownish color. This may be so slight as to be hardly preceptible, or it may be as dark as that of weak tea. It is darkest in water draining from swamps, and the color of the water in any pond or stream bears a close relation to the amount of swamp-land upon the tributary watershed.

The color is due to dissolved substances of vegetable origin extracted from leaves, peaty matter, etc. It is quite as harmless as tea. The exact chemical nature of the coloring matter is not known. It is complex in composition. Tannins, glucosides, and their derivatives are doubtless present. The color of a water usually bears a close relation to the albuminoid ammonia present. Carbon, however, is the important element in its composition. The color of a water varies very closely with the "oxygen consumed." Iron is usually present, and its amount varies with the depth of the color. In some waters iron alone imparts a high color, but in peaty waters it plays a subsidiary part.

The color of a water is usually stated in figures based on comparisons made with some arbitrary standard, the figures increasing with the depth of the color. The Platinum-Cobalt Standard, the Natural Water Standard, and the Nessler Standard are those most commonly used. The former gives the most satisfactory results. Comparisons of the water with the standard may be made in tall glass tubes or in a calorimeter such as that used at the Boston Water Works.*

The amount of color in the water collected from a watershed has a seasonal variation. This may be illustrated by the color of the water in Cold Spring Brook, at the head of Basin No. 4, Boston Water Works. This brook is fed in part from several large swamps. The figures given are based on weekly observations.

AVERAGE COLOR OF WATER IN COLD SPRING BROOK, 1894.

Jan.	Feb.	Mar.	Apr.	May.	June.	July.	Aug.	Sept.	Oct.	Nov.	Dec.	Av.
.99	.88	.96	.93	1.42	1.59	.98	.75	.60	.69	1.44	1.20	1.04

---

* See FitzGerald and Foss, "On the Color of Water," Jour. Frank. Inst., Dec. 1894.

There are usually two well-defined maxima, one in May or June and one in November or December. In the winter and early spring the color of the water is low because of dilution by the melted snow. As the yield of the watershed diminishes the color increases until the water standing in the swamp areas ceases to be discharged into the stream. During the summer the water in the swamps is high-colored, but its effect is not felt in the stream until the swamps overflow in the fall. Heavy rains during the summer may cause the swamps to discharge and increase the color of the water in the reservoirs below. It has been found that in general the color of the water delivered from any watershed bears a close relation to the rainfall. In some localities this is more noticeable than in others. In Massapequa Pond of the Brooklyn water-supply the color varies greatly from week to week, and the fluctuations are almost exactly proportional to the rainfall. In large bodies of water the seasonal fluctuations in color are less pronounced.

The hue of the water in the autumn is somewhat different from that in the spring. The fresh-fallen leaves and vegetable matter give a greenish-brown color that is quite different from the reddish-brown color produced from old peat.

When colored water is exposed to the light it becomes bleached. An elaborate series of experiments made at the Boston Water Works by exposing bottles of high-colored water to direct sunlight for known periods showed that during 100 hours of bright sunlight the color was reduced about 20%, and that with sufficient exposure all the color might be removed. The bleaching action was found to be independent of temperature. Sedimentation had but little influence on it. It was dependent entirely upon the amount of sunlight. The percentage reduction was independent of the original color of the water.

This bleaching action takes place in reservoirs where colored water is stored. Stearns has stated that in an unused reservoir 20 ft. deep the color of the water decreased from .40 to .10 in six months. In Basin No. 4, referred to above, the average color of the water in the influent stream for the year 1894 was 1.04. For the same year the average color of the water at the lower end of the basin was .71. It should be stated that this difference is not due wholly to bleaching action. The amount of coloring-matter entering the reservoir is not shown by the figure 1.04, for the reason that the quantity of water flowing in the stream is not uniform. It is greatest in the spring when the melting snows give the water a color lower than the average. Furthermore, some colorless rain-water and ground-water enters the basin. There is also a loss of high-colored water at the wasteway at a season when the color of the water is above the average. It is a difficult matter to ascertain just the amount of bleaching action that takes place in a reservoir through which water is constantly flowing.

Experiments (by the author) made by exposing bottles of colored water at various depths in reservoirs have shown that the bleaching action that takes place at the surface of a reservoir is considerable,—sometimes 50% in a month. It decreases rapidly with increasing depth, and the rapidity with which it decreases below the surface depends upon the color of the water in the reservoir, as the table on the following page will show.

From these and many similar experiments it has been found possible to calculate the extent of the bleaching action that takes place in any reservoir. The results agree closely with the observed color-readings of the water in the reservoir. The experiments also bear directly upon the point under discussion, namely, the penetration of light into the water of a reservoir.

EXPERIMENTS TO DETERMINE THE AMOUNT OF BLEACHING ACTION AT DIFFERENT DEPTHS.

	Expt. No. 1.	Expt. No. 2.
Color of water in reservoir	.20	.44
Time of exposure	Aug. 6–Sept. 4	July 2–Aug. 3
Color of water exposed	1.75	1.70
Percentage reduction of color:		
At depth of 0.50 ft.	65%	20%
" " " 1.25 "	32%	12%
" " " 2.5 "	21%	4%
" " " 5.0 "	14%	3%
" " " 7.5 "	3%	0%
" " " 10.0 "	1%	0%
" " " 15.0 "	0%	0%

Turbidity of Water.—Turbidity of water is due to the presence of particles of matter in suspension. These particles sometimes give color to water in mass, but this color is to be distinguished from that due to substances in solution.

It is difficult to estimate the amount of turbidity of water. Diaphanometers have been designed, but none of them are quite satisfactory. They measure the transparency of water by ascertaining through what depth of liquid a black-and-white figure can be seen with a given intensity of light.

A convenient method of comparing the turbidities of different waters that are very turbid is by means of the visibility of a platinum wire lowered horizontally in the water. This method was used first by Allen Hazen at the Lawrence Experiment Station. A bright platinum wire one millimeter in diameter was made to project at right angles from the lower end of a graduated rod. This was lowered into the water until the wire became invisible, when the depth of the wire was noted. A water that had such a turbidity that the wire became invisible at a depth of one inch was assumed to have a turbidity of one degree. Other degrees of turbidity were obtained by dividing one by the depth of the wire in inches. That is, a water in which the wire disappeared from view at a depth of 5 inches had a turbidity of 0.2; a water in which the wire disappeared from view at a depth of 20 inches had a turbidity of .05, etc. With comparatively clear waters the use of the black-and-white disc, as suggested by the author, will be found more satisfactory than that of the platinum wire.

Perhaps the most complete studies of the transparency of water on record are those made by Forel and others in Switzerland. Three methods of experiment were employed. The first was that of the visibility of plates. This method, used by Secchi in 1865 in determining the transparency of the water in the Mediterranean Sea, consisted of lowering a white disc (20 cm. in diameter) into the water and noting the depth at which it disappeared from view, and then raising it and noting the point at which it reappeared. The mean of these two depths was called the limit of visibility. The second method, known as that of the Genevan Commission, was similar to the first, but instead of a white disc an incandescent lamp was lowered into the water. This light when seen through the water from above presented an appearance similar to that of a street-lamp in a fog; that is, there was a bright spot surrounded by a halo of diffused light. When the light was lowered into the water the bright spot first disappeared from view. The depth of this point was noted as the "limit of clear vision." Finally the diffused light disappeared, and the depth of this point was called the "limit of diffused light." Both these methods were useful only in comparing the relative transparency of different waters or of the same water at different times. In order to get an idea of the intensity of light at different depths a photographic method was used. Sheets of sensitized albumen paper were mounted in a frame in such a way that half of the sheet was covered with a black screen, while the other half was exposed. A series of these papers was attached to a rope and lowered into the water; they were equidistant and so supported that they assumed a horizontal position in the water. They were placed in position in the night and allowed to remain 24 hours. On the next night they were drawn up and placed in a toning-bath. A comparison of prints made at different depths enabled the observer to determine the depth at which the light ceased to affect the paper and to obtain an idea of the relative intensity of the light at different depths. To assist in this comparison an arbitrary scale was made by exposing sheets of the same paper to bright sunlight for different lengths of time.

The results of the experiments are given by Forel as follows:

In Lake Geneva the limit of visibility of a white disk 20 cm. in diameter was 21 m. The limit of clear vision of a 7-candle-power incandescent lamp was 40 m.; the limit of diffused light was about 90 m. The depth at which the light ceased to affect the photographic paper was 100 m., when the paper was sensitized with chloride of silver, and about 200 m. when sensitized with iodobromide of silver. These depths were less in summer than in winter on account of the increased turbidity of the water. The transparency of the water in other lakes, as shown by the limit of visibility of a white disk, is cited as follows: Lake Tahoe, 33 m.; La Mer des Antilles, 50 m.; Lac Lucal, 60 m.; Mediterranean Sea, 42.5 m.;

Pacific Ocean, 59 m. It should be remembered that these are all comparatively clear and light-colored waters, and that in them the light penetrates to far greater detph than in turbid and colored water. For example, in Chestnut Hill Reservoir, a disc lowered into the water at a time when the color was 0.92 disappeared from view at a depth of six feet.

The author's experiments have shown that the limit of visibility may be determined most accurately by using a disc about 8 inches in diameter, divided into quadrants painted alternately black and white like the target of a level-rod, and looking vertically down upon it through a water-telescope provided with a suitable sunshade. It has been found that the limit of visibility obtained in this manner bears a close relation to the turbidity of the water as determined by a diaphanometer. It also varies with the color of the water, but the relation has not been carefully worked out.

Absorption of Light by Water.—The absorption of light by distilled water is said to vary with the temperature. The following coefficients are given by Wild as the result of laboratory experiments. It seems probable that the figures are too low.

Temperature.	Intensity of Light after passing through 1 dm. of Distilled Water.
24.4° C.	0.9179
17.0	0.93968
6.2	0.94769

The coefficient of absorption of light by colored water is quite unknown.

The reduction of light in passing downward through a body of water is supposed to follow the law that as the depth increases arithmetically the intensity of the light decreases geometrically. For example, if the intensity of the light falling upon the surface of a pond is represented by 1, and if $\frac{1}{4}$ of the light is absorbed by the first foot of water (some colored waters absorb even more than this), then the intensity of light at the depth of 1 ft. will be $\frac{3}{4}$; the second foot of water will absorb $\frac{1}{4}$ of $\frac{3}{4}$, and the intensity at the depth of 2 ft. will be $\frac{9}{16}$; and so on. At this rate of decrease the intensity of light at a depth of 10 ft. will be only about 5% of that at the surface.

There are few accurate data extant regarding the quality of the light at different depths, but theory would lead us to infer that in passing downward from the surface to the bottom of a lake the light varies considerably in character. It is said that the red and yellow rays are most readily transmitted.

CHAPTER VI.

GEOGRAPHICAL DISTRIBUTION OF MICROSCOPIC ORGANISMS IN PONDS AND LAKES.

THE microscopic organisms that are found most commonly in water-supplies taken from lakes or storage reservoirs are given in the following table,* arranged according to the usual system of classification and divided into groups according to their abundance and frequency of occurrence. The first group includes those genera which, in their season, are often found in large numbers; the second group includes those which are found but occasionally in large numbers; the third, those which often occur in small numbers; the fourth, those which are rarely observed. This division, while not wholly satisfactory, enables one to separate the important from the unimportant forms. As observations multiply, the list may be extended and some genera may be changed from one group to another. The organisms printed in heavy type have given trouble in water-supplies, either by producing odors or by making the water turbid and unsuitable for laundry purposes.

DIATOMACEÆ.

Commonly found in large numbers. **Asterionella,** Cyclotella, **Melosira, Synedra, Tabellaria.**

Occasionally found in large numbers. Diatoma, Fragilaria, Nitzschia, **Stephanodiscus.**

---

* Compiled from published biological examinations of Massachusetts water-supplies.

Commonly found in small numbers. Epithemia, Gomphonema, Navicula, Stauroneis.

Occasionally observed. Achnanthes, Amphiprora, Amphora, Bacillaria, Cocconeis, Cocconema, Cymbella, Diadesmis, Encyonema, Eunotia, Grammatophora, Himantidium, Isthmia, **Meridion,** Odontidium, Orthosira, Pinnularia, Pleurosigma, Schizonema, Striatella, Surirella, Tetracyclus.

CHLOROPHYCEÆ.

Commonly found in large numbers. Chlorococcus, Protococcus, **Scenedesmus.**

Occasionally found in large numbers. Cœlastrum, Cosmarium, **Palmella, Pandorina,** Polyedrium, Raphidium, Staurastrum, **Volvox.**

Commonly found in small numbers. Closterium, Conferva, Desmidium, Euastrum, **Eudorina,** Gonium, Micrasterias, Ophiocytium, Pediastrum, Sphærozosma, Staurogenia, Tetraspora, Ulothrix, Xanthidium.

Occasionally observed. Arthrodesmus, Bambusina, Botryococcus, Characium, Chætophora, Cladophora, Dactylococcus, Dictyosphærium, Dimorphococcus, Draparnaldia, Glœocystis, Hyalotheca, Mesocarpus, Nephrocytium, Penium, Selenastrum, Sorastrum, Spirogyra, Stigeoclonium, Tetmemorus, Zygnema.

CYANOPHYCEÆ.

Commonly found in large numbers. **Anabæna, Clathrocystis, Cœlosphærium, Microcystis.**

Occasionally found in large numbers. **Aphanizomenon,** Chroöcoccus, **Oscillaria.**

Commonly found in small numbers. Aphanocapsa.

Occasionally observed. Glœocapsa, Lyngbya, Merismopedia, Microcoleus, Nostoc, Rivularia, Sirosiphon, Tetrapedia.

GEOGRAPHICAL DISTRIBUTION OF ORGANISMS. 79

SCHIZOMYCETES AND FUNGI.

Commonly found in large numbers. **Crenothrix.**

Occasionally found in large numbers. Cladothrix.

Commonly found in small numbers. **Beggiatoa,** Leptothrix, Molds.

Occasionally observed. Achlya, Leptomitus, Saprolegnia, Sarcina, Spirillum.

PROTOZOA.

Commonly found in large numbers. **Cryptomonas, Dinobryon, Peridinium, Synura, Uroglena.**

Occasionally found in large numbers. **Bursaria,** Chloromonas, **Glenodinium, Mallomonas, Raphidomonas.**

Commonly found in small numbers. Actinophrys, Amœba, Anthophysa, Ceratium, Cercomonas, Codonella, Epistylis, Monas, Tintinnus, Trachelomonas, Vorticella.

Occasionally observed. Acineta, Arcella, Chlamydomonas, Coleps, Colpidium, Cyphodera, Difflugia, Enchelys, Euglena, Euglypha, Euplotes, Glaucoma, Halteria, Heteronema, Nassula, Paramæcium, Phacus, Pleuronema, Raphidodendron, Stentor, Syncrypta, Trichodina, Uvella, Zoothamnium.

ROTIFERA.

Commonly found in small numbers. Anuræa, Conochilus, Polyarthra, Rotifera, Synchæta.

Occasionally observed. Asplanchna, Colurus, Eosphora, Floscularia, Lacinularia, Mastigocerca, Microcodon, Monocerca, Monostyla, Noteus, Sacculus, Triarthra.

CRUSTACEA.

Commonly found in small numbers. Bosmina, Cyclops, Daphnia.

Occasionally observed. Alona, Cypris, Diaptomus, Sida.

MISCELLANEOUS.

Occasionally observed. Acarina, Anguillula, Batrachospermum, Chætonotus, Gordius, Hydra, Macrobiotus, Meyenia, Nais, Spongilla; besides spores, ova, insect scales, pollen-grains, vegetable fibres and tissue, yeast-cells, starch-grains, etc.

The above may be summarized numerically as follows:

Classification.	Number of Genera.				
	Commonly found in large numbers.	Occasionally found in large numbers.	Commonly found in small numbers.	Occasionally observed.	Total.
Diatomaceæ	5	4	4	22	35
Chlorophyceæ............	3	8	14	21	46
Cyanophyceæ............	4	3	1	8	16
Fungi and Schizomycetes	1	1	3	5	10
Protozoa	5	5	11	24	45
Rotifera.................	0	0	5	12	17
Crustacea	0	0	3	4	7
Miscellaneous...........	0	0	0	10	10
Total	18	21	41	106	186

It will be observed that 186 genera have been recorded,—108 plants and 78 animals. Of these only 18 are commonly found in large numbers,—13 plants and 5 animals. 21 more are occasionally found in large numbers,—16 plants and 5 animals. 41 genera are frequently seen in small numbers, while 106 genera, or more than one half of all are seen occasionally, some of them rarely. The most, important classes are the Diatomaceæ, Chlorophyceæ, Cyanophyceæ, and Protozoa, as shown by the large number of genera and by their greater abundance. Furthermore, these classes include all but one of the most troublesome genera that have been found in large numbers. 10 genera may be said to be very troublesome because of their wide distribution, the frequency of their occurrence, and their unpleasant effects. They are Asterionella, Anabæna, Clathrocystis, Cœlosphærium, Aphanizomenon,* Dinobryon, Peridinium, Synura, Uroglena, and Glenodinium. This list seems like a short one when one considers the annoyance that the microscopic organisms have caused in various water-supplies.

The observations of sanitarians and the planktologists show that the microscopic organisms are very widely distributed in nature. They are found in all parts of the world, and under great varieties of climatic conditions. It is probable that they appeared on the earth at an early geological age. Some of them are found as fossils,—notably the diatoms, which have silicious walls that are almost indestructible.

In spite of the vast amount of study that has been given to the microscopic organisms we are still very far from understanding the laws governing their distribution. Why it is that a certain genus will grow vigorously in one pond and at the same time be absent from a neighboring one where the conditions apparently are as favorable, or why a form may suddenly appear in a pond where it has been never before seen, we are still unable to say with certainty. Solution of such problems involves a far-reaching knowledge of the chemical constituents and the life-history of the organisms, besides the effect of physical conditions, such as temperature, pressure, light, etc. The sciences of bio-chemistry and bio-physics are yet in their infancy. Until these have been further developed many problems connected with the microscopic organisms must remain unsolved.

The following statistics are of some value in connection

---

* In the reports of the Massachusetts State Board of Health this organism is sometimes classed with Oscillaria.

CLASSIFICATION OF FIFTY-SEVEN MASSACHUSETTS PONDS AND RESERVOIRS.

(The figures refer to the classes to which the ponds belong.)

City or Town	Pond or Reservoir	Depth	Diatomaceæ	Chlorophyceæ	Cyanophyceæ	Protozoa	Color	Excess of Chlorine	Hardness	Albuminoid Ammonia (in solution)	Free Ammonia	Nitrates
Abington	Big Sandy Pond	0	III	IV	IV	III	I	III	I	II	I	II
Andover	Haggett's Pond	0	III	II	IV	II	I	I	I	III	I	I
Ashburnham	Upper Naukeag Pond	0	IV	II	IV	III	I	I	II	II	I	II
Athol	Phillipston Reservoir	0	III	II	IV	III	I	I	II	II	I	I
Boston	Reservoir No. 6	+	III	III	IV	IV	II	III	III	IV	III	II
Boston	Reservoir No. 4	+	III	III	III	III	III	II	III	IV	III	III
Boston	Reservoir No. 2	0	III	III	III	II	III	III	IV	IV	III	III
Boston	Reservoir No. 3	0	I	I	II	III	III	III	IV	IV	III	IV
Boston	Lake Cochituate	+	I	I	II	I	I	IV	IV	III	III	IV
Boston	Farm Pond	0	I	I	II	I	I	IV	IV	III	III	IV
Boston	Mystic Lake	+	III	III	IV	III	I	IV	III	IV	III	III
Boston	Jamaica Pond	+	I	II	II	III	II	II	III	III	I	II
Brockton	Salisbury Brook Reservoir	0	III	III	IV	III	I	III	III	III	I	I
Cambridge	Fresh Pond	+	III	III	IV	II	III	IV	III	III	I	II
Cambridge	Stony Brook Reservoir	0	III	III	IV	III	III	IV	III	III	III	I
Danvers	Middleton Pond	+	I	I	II	II	II	I	I	I	III	I
Fitchburg	Scott Reservoir	÷	III	III	III	III	I	II	I	III	I	II
Fitchburg	Falulah Reservoir	0	IV	IV	IV	IV	I	I	I	IV	IV	I
Fitchburg	Meeting House Pond	0	III	III	III	II	I	III	II	IV	I	I
Gardner	Crystal Lake	0	II	II	II	I	I	I	II	I	I	II
Gloucester	Dikes Brook Reservoir	+	I	I	II	I	I	I	II	III	I	II
Gloucester	Wallace Pond	0	II	II	II	I	II	II	I	IV	IV	III
Gloucester	Fernwood Lake	0	I	I	II	III	IV	III	I	I	I	II
Hingham	Accord Pond	0	I	I	III	I	I	I	I	I	I	II
Hinsdale	Fulling Mill Pond	0	II	II	II	III	I	II	II	IV	III	III
Hinsdale	Storage Reservoir	+	II	II	II	I	III	III	IV	IV	IV	III
Holyoke	Whiting Street Reservoir	0	I	I	IV	I	I	I	III	I	I	I
Holyoke	Ashley Pond	0	I	IV	IV	III	I	II	IV	III	II	II

GEOGRAPHICAL DISTRIBUTION OF ORGANISMS.

Location	Depth										
Hudson	Gates Pond	+	III	—	III	III	—	III	II	II	II
Leominster	Haynes Reservoir	O	—	II	II	II	III	II	—	IV	II
Lynn	Breeds Pond	O	—	III	III	III	III	II	II	IV	III
Lynn	Birch Pond	O	—	—	—	—	—	III	III	III	III
Lynn	Walden Pond	O	—	III	—	—	—	IV	III	III	III
Lynn	Glen Lewis Pond	O	—	—	III	III	III	III	IV	III	III
Malden	Spot Pond	+	II	II	III	—	—	IV	III	III	III
Marlboro	Lake Williams	O	III	III	II	III	III	—	IV	III	III
Montague	Lake Pleasant	O	III	III	III	—	—	IV	IV	IV	III
Nantucket	Wannacomet Pond	O	III	II	III	II	—	II	III	III	III
Natick	Dug Pond	+	IV	IV	IV	—	IV	II	III	II	II
New Bedford	Lake Quittacus	O	II	II	II	III	II	IV	III	II	II
North Brookfield	Doane Pond	O	III	III	III	III	II	III	II	III	III
Norwood	Buckmaster Pond	O	—	III	II	IV	—	III	II	III	—
Orange	North Pond	O	—	II	IV	IV	—	II	IV	III	II
Plymouth	Little South Pond	O	—	II	III	III	II	IV	III	—	II
Quincy	Storage Reservoir	O	—	—	II	II	IV	—	II	II	—
Rockport	Cape Pond	O	—	IV	IV	—	—	IV	III	II	II
Salem	Wenham Lake	O	—	IV	IV	—	II	II	IV	IV	—
Springfield	Ludlow Reservoir	O	III	IV	II	II	III	III	III	III	III
Taunton	Elders Pond	O	III	IV	IV	III	II	III	II	III	III
Westboro	Upper Sundra Reservoir	O	II	II	IV	—	II	III	II	II	—
Winchester	North Reservoir	+	III	III	—	—	IV	—	III	III	III
Winchester	South Reservoir	O	I	—	I	—	—	I	—	—	—
Winchester	Middle Reservoir	O	III	III	I	II	II	III	III	III	III
Woburn	Horn Pond	+	—	II	I	III	II	—	I	IV	IV
Worcester	Lynde Brook Reservoir	O	II	II	II	—	III	—	IV	IV	II
Worcester	Tatnuck Brook Reservoir	O	III	III	III	III	III	III	III	II	II

DEPTH. + Deep. O Shallow.

ORGANISMS. I. Often above 1000 per c.c.
II. Occasionally above 1000 per c.c.
III. Usually from 100 to 500 per c.c.
IV. Below 100 per c.c.

EXCESS OF CHLORINE. I. 0
II. .01 to .03
III. .04 to .25
IV. Above .25

HARDNESS. I. 0 to 0.5
II. 5 to 1.0
III. 1.0 to 2.0
IV. Above 2.0

NITRATES. I. 0 to .0050
II. .0050 to .0100
III. .0100 to .0200
IV. Above .0200

FREE AMMONIA. I. 0 to .0010
II. .0010 to .0030
III. .0030 to .0100
IV. Above .0100

ALBUMINOID AMMONIA (dissolved). I. 0 to .0100
II. .0100 to .0150
III. .0150 to .0200
IV. Above .0200

COLOR. I. 0 to 0.30 Nessler Scale.
II. 0.30 to 0.60 Nessler Scale.
III. 0.60 to 1.00 Nessler Scale.
IV. Above 1.00 Nessler Scale.

The figures of the chemical analyses are given in parts per 100,000.

with this subject, as they show the relative abundance of the different classes of organisms in some of the important surface-water supplies of Massachusetts, together with some of the elements of the sanitary chemical analysis.

For the purpose of this comparison 57 ponds and reservoirs were selected where monthly examinations, both chemical and biological, have been carried on for a number of years by the State Board of Health. The results of these examinations were carefully studied, and the ponds (which, for convenience, we may consider to include lakes, ponds, and storage reservoirs) divided into groups as shown in the table on pages 82 and 83.

The first two columns in this table give the names of the ponds and the cities which they supply. The third gives the depth of the pond, whether shallow or deep. The next four columns show the relative abundance of the four most important classes of organisms; namely, the Diatomaceæ, Chlorophyceæ, Cyanophyceæ, and Protozoa. The four groups are characterized as follows: the group to which each pond belongs is indicated by a Roman numeral.

Group I. Number of organisms often as high as 1000 per c.c.

Group II. Number of organisms only occasionally as high as 1000 per c.c.

Group III. Number of organisms ordinarily between 100 and 500 per c.c.

Group IV. Number of organisms never above 100 per c.c.

These figures refer not to the numbers present in the average sample of water, but to the numbers during the season of maximum growth. The boundaries of the groups were not sharply defined, and in a number of cases it was hard to tell whether a pond should be classed in group II or

III. The last five columns show the ponds divided into classes according to some of the elements of the chemical analysis; namely, color, excess of chlorine, hardness, albuminoid ammonia (in solution), free ammonia, and nitrates In each case four classes are given, division being made according to the schedule given at the bottom of the table.

If we consider the ponds with reference to the growths of organisms, we obtain from the above table the following summary:

Group.	Number per c.c.	Number of Ponds and Reservoirs.			
		Diatomaceæ.	Chlorophyceæ.	Cyanophyceæ.	Protozoa.
I	Often above 1000 per c.c............	24	5	7	8
II	Occasionally above 1000 per c.c.....	8	11	10	7
III	Usually between 100 and 500 per c.c..	19	29	18	35
IV	Below 100 per c.c.....	6	12	22	7

From this it appears that the Diatomaceæ are the organisms most commonly found in large numbers. There are 24 ponds (42%) of the ponds considered) which often have these organisms as high as 1000 per c.c., while in only 6 (11%) are they always below 100 per c.c. The Chlorophyceæ are not often found in great abundance, though many ponds contain them in moderate numbers. Only 5 ponds (9%) have growths of 1000 per c.c., while 29 (70%) have growths of from 100 to 500 per c.c. The Cyanophyceæ are not as common as the Chlorophyceæ, but where they do occur their growth is usually greater and they cause more trouble. There are 7 ponds (12%) that commonly have growths above 1000 per c.c., while in 22 (39%) they are never above 100 per c.c. The Protozoa are somewhat more abundant than either the Chlorophyceæ or Cyanophyceæ. Eight ponds (12%) often have growths above 1000 per c.c.; 35 ponds (60%) have growths between 100 and 500 per c.c.

From the table on pages 82 and 83 it also appears that 28 ponds (49%) often have high growths of one or more of these classes of organisms at some time during the year. Such growths, except in the case of certain diatoms, are nearly always noticeable and frequently are very troublesome. In 17 ponds the Diatomaceæ alone reach 1000 per c.c.; in 1 pond the Cyanophyceæ alone; and in 3 ponds the Protozoa alone. One pond has heavy growths of Diatomaceæ, Chlorophyceæ and Protozoa; two, of Diatomaceæ, Chlorophyceæ and Cyanophyceæ; two, of Diatomaceæ, Cyanophyceæ and Protozoa. In two ponds all four classes are found in large numbers. There is but one pond where the organisms never rise above 100 per c.c.; there are 16 where no class of organisms shows numbers greater than 500 per c.c.

For the purpose of determining whether the depth of the pond exercises any important influence upon the growth of the organisms the following table was compiled:

Depth.*	Number per c.c.	Number of Ponds.			
		Diatomaceæ.	Chlorophyceæ.	Cyanophyceæ.	Protozoa.
Deep.....	Often above 1000 per c.c............	8	3	2	2
Deep.....	Occasionally above 1000 per c.c........	2	2	1	0
Deep.....	Usually between 100 and 500 per c.c....	6	8	6	12
Deep.....	Always below 100 per c.c.	0	3	7	2
Shallow..	Often above 1000 per c.c............	16	2	5	6
Shallow..	Occasionally above 1000 per c.c.........	6	9	9	7
Shallow..	Usually between 100 and 500 per c.c.....	13	21	12	23
Shallow..	Always below 100 per c.c.............	6	9	15	5

* Ponds of the Second Order are here called "deep ponds"; ponds of the Third Order "shallow ponds"; no ponds of the First Order are included. See page 63.

There are 16 deep and 41 shallow ponds. Of the deep ponds 63% at times have growths of the Diatomaceæ above 1000 per c.c., while of the shallow ponds 54% have such growths. There are no deep ponds where the Diatomaceæ are lower than 100 per c.c., while in 15% of the shallow ponds they are lower than that figure. It thus appears that the heavy growths of the Diatomaceæ are somewhat more likely to be found in the deep than in the shallow ponds. The same may be said of the Chlorophyceæ, though the difference is not so marked. 31% of the deep ponds and 27% of the shallow ponds at times have growths as high as 1000 per c.c. The Cyanophyceæ and Protozoa, on the other hand, incline toward shallower water. In the case of the former, 18% of the deep ponds and 34% of the shallow ponds at times have growths of 1000 per c.c., while in the case of the latter the figures are 12% and 32% respectively

In this connection it would be of interest to show statistically the relation that undoubtedly exists between the growths of organisms and the character of the material forming the bottoms of the ponds, but unfortunately the necessary data are lacking in too many cases. So far as observations have been made, it appears that muddy bottoms are very largely responsible for excessive growths of microscopic organisms.

An important question, and one which is of particular interest to water analysts, is the relation between the growths of organisms and the chemical analysis of the water in which the organisms are found. Unquestionably there is such a relation, and we should very much like to be able to take up a chemical analysis and say "this water contains such and such substances in solution, and, therefore, such and such organisms may be expected to thrive well in it." In other words, we desire to know better the nature of the necessary food-supply of the microscopic organisms.*

---

* Experiments upon this subject are now in progress.

The tables given on pages 89 and 90 are designed to show in a very general way the relation between the organisms in the 57 selected ponds and some of the important elements of the chemical analysis.

These tables reveal several important facts: first, it is seen that the color of a water has an important influence upon the number of organisms that will be found in it. Of the 24 cases where the Diatomaceæ are commonly found higher than 1000 per c.c., 12 (or 50%) occur in light-colored waters, i.e., water having a color lower than 0.30 on the Nessler scale, and none occur in water where the average color is above 1.00. The same fact is noticed in the case of the other organisms, but not as markedly as with the Diatomaceæ. The reason for this is, doubtless, on account of the difference in specific gravity between the diatoms and the other organisms. The diatoms are heavy by reason of their siliceous cell-walls, but the other organisms are much lighter and some of them liberate gas, causing them to keep near the surface. The depth to which light penetrates in a body of water makes less difference with the growth of the Cyanophyceæ, for example, than it does with the diatoms, which constantly tend to sink and which are kept near the surface chiefly by the vertical currents in the water.

The "excess of chlorine" means the difference between the amount of chlorine found in a sample of water and that found in the unpolluted water of the same region. To a certain extent it represents the amount of pollution which the water has received. It is important to know whether this element of the analysis bears any relation to the organisms and whether one may rightly infer that a large growth of organisms in a reservoir is any indication of the pollution of a water-supply. A study of the tables shows that only to a

GEOGRAPHICAL DISTRIBUTION OF ORGANISMS. 89

A.

Chemical Analysis (parts per 100,000).		Number of Ponds and Reservoirs in which the Diatomaceæ are			
		Often above 1000 per c.c.	Occasionally above 1000 per c.c.	Usually between 100 and 500 per c.c.	Below 100 per c.c.
Color (Nessler Scale)	0 to .30	12	4	9	4
	.30 to .60	6	2	4	0
	.60 to 1.00	6	1	5	1
	above 1.00	0	1	1	1
Excess of Chlorine	0	4	2	1	2
	.01 to .03	8	1	8	2
	.04 to .25	8	3	10	2
	above .25	4	2	0	0
Hardness	0 to .5	2	1	3	3
	.5 to 1.0	7	4	5	2
	1.0 to 2.0	8	0	10	1
	above 2.0	7	3	1	0
Albuminoid Ammonia (dissolved)	0 to .0100	2	0	2	1
	.0100 to .0150	6	1	5	3
	.0150 to .0200	8	6	7	1
	above .0200	8	1	5	1
Free Ammonia	.0000 to .0010	3	2	5	3
	.0010 to .0030	6	1	10	2
	.0030 to .0100	8	5	4	1
	above .0100	7	0	0	0
Nitrates	0 to .0050	3	3	5	6
	.0050 to .0100	11	3	13	0
	.0100 to .2000	6	2	1	0
	above .0200	4	1	0	0

B.

Chemical Analysis (parts per 100,000).		Number of Ponds and Reservoirs in which the Chlorophyceæ are			
		Often above 1000 per c.c.	Occasionally above 1000 per c.c.	Usually between 100 and 500 per c.c.	Below 100 per c.c.
Color (Nessler Scale)	0 to .30	2	5	14	8
	.30 to .60	2	4	5	1
	.60 to 1.00	1	2	8	2
	above 1.00	0	0	2	1
Excess of Chlorine	0	1	3	4	1
	.01 to .03	1	2	11	5
	.04 to .25	0	4	13	6
	above .25	3	2	1	0
Hardness	0 to .5	0	2	3	4
	.5 to 1.0	1	4	8	5
	1.0 to 2.0	1	3	13	2
	above 2.0	3	2	5	1
Albuminoid Ammonia (dissolved)	0 to .0100	0	0	2	3
	.0100 to .0150	0	4	7	4
	.0150 to .0200	2	5	12	3
	above .0200	3	2	8	2
Free Ammonia	0 to .0010	0	2	7	4
	.0010 to .0030	0	1	13	5
	.0030 to .0100	2	5	8	3
	above .0100	3	3	1	0
Nitrates	0 to .0050	0	2	8	6
	.0050 to .0100	2	6	13	6
	.0100 to .0200	0	2	7	0
	above .0200	3	1	1	0

C.

Chemical Analysis (parts per 100,000).		Number of Ponds and Reservoirs in which the Cyanophyceæ are			
		Often above 1000 per c.c.	Occasionally above 1000 per c.c.	Usually between 100 and 500 per c.c.	Below 100 per c.c.
Color (Nessler Scale)	0 to .30	2	4	12	11
	.30 to .60	2	3	4	3
	.60 to 1.00	3	2	1	7
	above 1.00	0	1	1	1
Excess of Chlorine	0	2	1	3	3
	.01 to .03	1	3	5	10
	.04 to .25	1	5	8	9
	above .25	3	1	2	0
Hardness	0 to .5	0	2	1	6
	.5 to 1.00	2	2	4	10
	1.00 to 2.00	2	5	7	5
	above 2.00	3	1	6	1
Albuminoid Ammonia (dissolved)	0 to .0100	0	0	1	4
	.0100 to .0150	0	3	6	6
	.0150 to .0200	2	5	8	7
	above .0200	5	2	3	5
Free Ammonia	0 to .0010	0	2	1	10
	.0010 to .0030	0	2	9	8
	.0030 to .0100	3	5	6	4
	above .0100	4	1	2	0
Nitrates	0 to .0050	1	2	1	12
	.0050 to .0100	3	4	10	10
	.0100 to .0200	1	3	5	0
	above .0200	2	1	2	0

D.

Chemical Analysis (parts per 100,000).		No. of Ponds and Reservoirs in which the Protozoa are			
		Often above 1000 per c.c.	Occasionally above 1000 per c.c.	Usually between 100 and 500 per c.c.	Below 100 per c.c.
Color (Nessler Scale)	0 to .30	5	2	20	2
	.30 to .60	1	3	6	2
	.60 to 1.00	2	2	8	1
	above 1.00	0	0	1	2
Excess of Chlorine	0	1	2	5	1
	.01 to .03	1	2	13	3
	.04 to .25	2	3	15	3
	above .25	3	0	3	0
Hardness	0 to .5	0	0	7	3
	.5 to 1.00	3	0	12	2
	1.0 to 2.00	1	6	10	2
	above 2.00	4	1	6	0
Albuminoid Ammonia (dissolved)	0 to .0100	0	0	4	1
	.0100 to .0150	0	0	13	2
	.0150 to .0200	5	2	12	3
	above .0200	3	4	7	1
Free Ammonia	0 to .0010	1	1	9	2
	.0010 to .0030	1	1	13	4
	.0030 to .0100	2	5	10	1
	above .0100	4	0	3	0
Nitrates	0 to .0050	0	1	12	3
	.0050 to .0100	3	4	17	3
	.0100 to .0200	3	2	3	1
	above .0200	2	0	3	0

small extent does the excess of chlorine influence the number of organisms observed, though there is a slight tendency for heavy growths of organisms to accompany high excess of chlorine. This fact corresponds with the common observation that vigorous growths of organisms are often observed in ponds far removed from any possible contamination.

The hardness of a water, i.e., the abundance of carbonates and sulphates of calcium and magnesium, appears to have some influence upon the organisms. This is noticed in all four classes, though it is most marked in the case of the Diatomaceæ and Protozoa. For example, of the 10 ponds low in hardness not one ever has the Protozoa as high as 1000 per c.c., while of the 11 ponds high in hardness every one has Protozoa above 100 per c.c., and 4 commonly have them above 1000 per c.c.

The sanitary chemical analysis ordinarily states the amount of nitrogen present in four different forms, namely, albuminoid ammonia (dissolved and suspended), free ammonia, nitrites, and nitrates, which represent four stages in the change of organic to inorganic matter. Since nitrogen is essential to all living matter we naturally expect that organisms will thrive best in waters rich in that element. The above statistics show that this is the case, and that it is true for each class of organisms and for the different conditions of nitrogen tabulated. The free ammonia and nitrates appear to be particularly influential in determining the amount of life present. For example, 10 of the 13 ponds low in free ammonia never show maximum growths of the Cyanophyceæ above 100 per c.c., while 4 of the 7 ponds high in free ammonia commonly have growths above 1000 per c.c.

One must be careful in these matters, however, not to mistake cause for effect. Free ammonia, for example, indicates organic matter in a state of decay, and instead of representing the food of the organisms in question it may represent their decomposition. The inter-action of the various organisms is a very complicated question, and the extent to which one organism lives upon the products of decay of another is not well known.

CHAPTER VII.

SEASONAL DISTRIBUTION OF MICROSCOPIC ORGANISMS.

THE microscopic organisms found in water show variations in their seasonal occurrence as great and almost as characteristic as do many land plants. The succession of dandelions, buttercups, and goldenrod in our fields finds its counterpart in the succession of the diatoms, the green algæ, and the blue-green algæ in our lakes and ponds. If one examines the water of a lake continuously for a year some interesting changes in its flora and fauna may be observed. If the lake is a typical one the water during the winter will contain comparatively few organisms: in the spring various diatoms will appear; these will disappear in a few weeks and in their place will come the green algæ: at the same time the blue-green algæ may be found; in the fall both of these will vanish and the diatoms will develop again; as the lake freezes these in turn will disappear. Similar but less characteristic fluctuations take place among the animal forms. These facts are shown graphically in Fig. 14, which represents the seasonal changes that occur among the more important organisms in Lake Cochituate. The diagram is based on weekly observations extending over a number of years. The seasonal distributions of the diatoms, algæ, etc., are so different that it is best to consider each class by itself.

Diatomaceæ.—In most natural ponds and storage reservoirs diatoms are far more abundant in the spring and fall than at other seasons. New growths seldom begin in the summer or winter, but the spring and fall growths sometimes linger into the summer and winter for a number of weeks.

The occurrence of diatoms in ponds is greatly influenced by the vertical circulation of the water. They generally

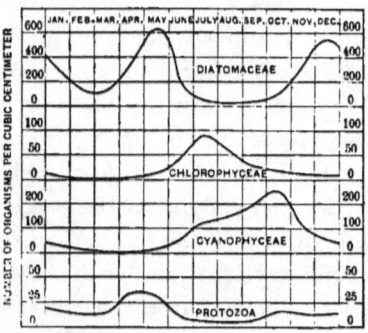

Fig. 14.—Seasonal Distribution of Microscopic Organisms in Lake Cochituate.

appear after the periods of stagnation and during the periods of complete vertical circulation. It has been found that in temperate lakes of the second order, which have well-marked periods of stagnation in summer and in winter, the spring and fall growths of Asterionella occur with great regularity and with about equal intensity, while in temperate lakes of the third order, which are stagnant only during the winter, the Asterionella growths in the autumn are either small compared with the spring growths or are lacking altogether. In deep ponds the spring growths occur earlier and the fall growths considerably later than in shallow ponds, thus again corresponding to the periods of circulation. In lakes of the third order diatoms are sometimes found during the summer after periods of partial stagnation.

Of the many genera of diatomaceæ that are observed in water only those that are true plankton forms exhibit the spring and fall maxima. The most important of these are Asterionella, Tabellaria, Melosira, Synedra, Stephanodiscus, Cyclotella, and Diatoma. Other genera are more uniformly distributed through the year. All of these seven genera are sometimes, but not often, observed in the same body of water. As a rule certain ponds have certain diatoms peculiar to them. For example, Lake Cochituate often contains large growths of Asterionella, Tabellaria, and Melosira; other diatoms are to be found, but they are seldom very numerous. Basin 3 of the Boston Water Works contains Asterionella, Tabellaria, and Synedra, but few Stephanodiscus or Melosira. In Basin 2 only Synedra and Cyclotella are found. In Basin 4 Cyclotella usually predominates. Fresh Pond, Cambridge, Mass., is famous for its Stephanodiscus, and Diatoma is common in the water-supply of Lynn, Mass.

The genera that appear in any pond are not the same every year. In Lake Cochituate the spring growth in 1890 consisted of Asterionella and Tabellaria; in 1891 of Asterionella with a few Melosira; in 1892 of Melosira chiefly; in 1893 of Melosira and Asterionella; and in 1894 of Tabellaria, Asterionella, and Melosira. Furthermore, in any season it is seldom that two genera attain their maximum development at the same time,—sometimes one appears first and sometimes another. The most interesting succession of genera that the author has observed occurred in 1892 in Chestnut Hill Reservoir of the Boston Water Works. The spring growth began in April and continued through July. For three months the total number of diatoms present did not materially change, but during this time six different genera appeared on the scene, culminated one after another, and disappeared. This is shown in Fig. 15.

The explanation of the peculiar seasonal distribution of diatoms involves the answers to many questions. To what extent are diatoms influenced by light, by temperature, by mechanical agitation? To what extent are they dependent upon oxygen or carbonic acid dissolved in water? What sort

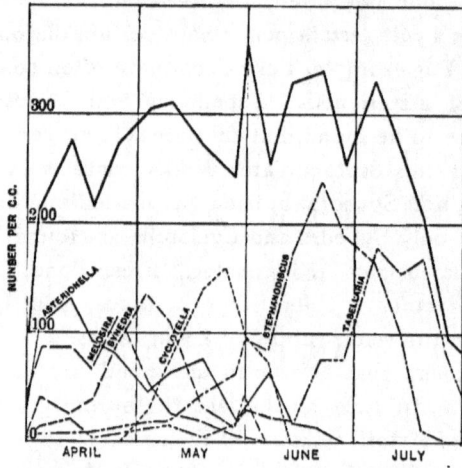

FIG. 15.—SUCCESSION OF DIATOMS IN CHESTNUT HILL RESERVOIR, 1892.

of mineral matter do they require? These are questions not yet fully answered. Attempts have been made to solve the problems by experiment, but it has been found difficult to control all the necessary conditions in the laboratory.

The optimum temperature for the development of the diatomaceæ is not known. Diatom growths have been observed at temperatures ranging from 35° to 75° F. In Lake Cochituate the average temperature of the water at the time of maximum Asterionella growths is not far from 50°.

In some lakes it is nearer 60°. Experimental evidence upon the subject is weak, but there is reason for believing that the optimum temperature for the diatomaceæ is lower than for the green or blue-green algæ.

It is known that diatoms are very sensitive to light. They will not grow in the dark nor in bright sunlight. Experiments* made by allowing diatoms to grow in bottles at various depths below the surface have shown that their growth is nearly proportional to the intensity of the light. This is

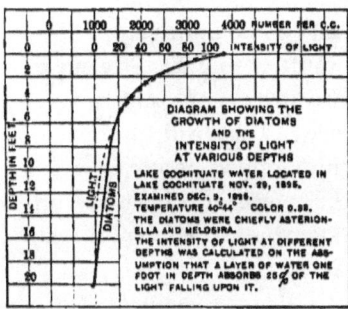

FIG. 16.

illustrated by Fig. 16. It will be noticed that near the surface,† where the light was strong, they multiplied rapidly, but below the surface the rate of multiplication was much slower, and at a certain depth no multiplication took place. This depth-limit of growth varied according to the color and transparency of the water, being greatest in the water having the least color. In one reservoir, where the color was 0.86, the limit of growth was 8 ft.; in another, where the color was

---

* By the author.

† The growth at the depth of 6 inches was greater than at the immediate surface, where the direct sunlight was too strong.

0.60, it was 12 ft.; and in another, with a color of 0.29, it was 15 ft. No observations were made in colorless waters, but in them the limit of growth is as great as 25 or 50 ft., and perhaps even much more than this.

The specific gravity of diatoms plays an important part in their seasonal distribution. In absolutely quiet water most diatoms sink to the bottom, but very slight vertical currents are sufficient to prevent them from sinking. A few forms appear to have a slight power of buoyancy, and some genera are somewhat motile.

Diatoms are said to be positively heliotropic, that is, they tend to move towards the light. In some of the motile forms this power is quite strong. In most of the plankton genera this power is weak. They will not move upwards towards the light through any great depth of water. It is possible, however, that the power of heliotropism varies with the intensity of the light, but experimental evidence on this point is lacking.

Diatoms require air for their best development. Experiment has shown that they will not multiply in a jar where a thin layer of oil covers the surface of the water: that in cultures in jars of various shapes, the one that has the least depth of water and the greatest amount of surface exposed to the air will show the greatest multiplication; that in bottles exposed at the same depth beneath the surface of a reservoir, one with bolting-cloth tied over the mouth will show a greater development of diatoms than one tightly stoppered.

The nature of the food-material of diatoms is not well known. Observations seem to show that they require nitrogen in the form of nitrates or free ammonia (perhaps both), silica, and more or less mineral matter, such as the salts of magnesium, calcium, iron, manganese, etc., but the amounts of these various substances required has not been determined.

These facts enable one to formulate a theory for the explanation of the occurrence of maximum growths of diatoms after the periods of stagnation and during the periods of circulation.

During the periods of stagnation the lower stratum of water in a deep lake undergoes certain changes that are very pronounced if the bottom of the lake holds any accumulation of organic matter. The organic matter decays, the oxygen becomes exhausted, decomposition proceeds under the action of the anaerobic bacteria, the free ammonia increases, and other organic and inorganic substances become dissolved in the water. During the period of circulation this foul water reaches the surface, further oxidation takes place, and compounds favorable to the growth of diatoms are formed. At the same time the vertical currents carry to the surface the diatoms, or their spores, that have been lying dormant at the bottom, where they could not grow because of darkness or because of the absence of proper food conditions. Carried thus towards the surface, where there is an abundance of light, air, and nutrition, they multiply rapidly. The extent of their development depends upon the amount of food-material present, the temperature of the water, and the amount of vertical circulation. If the upper layers become stratified and the surface remains calm for a number of days the diatoms will settle in the water into a region where the light is less intense. If they sink far enough they enter a region where the light is not sufficient for their growth, and if they sink below the thermocline succeeding vertical circulation of the upper strata will not affect them. Unable to reach the surface by their own power they will sink to the bottom and remain through another period of stagnation.

In small reservoirs that are constantly supplied with water rich in diatom food and that are so shallow that even at the bottom the light is strong enough for their development, the seasonal distribution follows somewhat different laws. This is the case in many open reservoirs where ground-water is stored.

Chlorophyceæ.—The Chlorophyceæ are most abundant in water-supplies during the summer. They are seldom found in winter. The curve showing their development (see Fig. 14) is more nearly parallel with the curve showing the temperature of the water than is that of any other class of organisms. The maximum growth is usually in July or August, though some genera culminate as early as June and others as late as September or even October. The late growths are usually associated with the phenomenon of stagnation.

The optimum temperature for the different genera is not known. It seems probable that any of the common forms are able to grow vigorously between 60° and 80° F. if their food-supply is favorable. It is possible for some of the green algæ to become acclimated to considerable extremes of heat or cold. Protococcus nivalis is found in the arctic regions, and Conferva has been observed in water at a temperature of 115° F.

Cyanophyceæ.—The seasonal distribution of the Cyanophyceæ is similar to that of the Chlorophyceæ, but as a rule the maximum growths occur a little later in the season. The Cyanophyceæ seem to be attuned to a slightly higher temperature than the Chlorophyceæ. They often show a great increase after a period of hot weather. Anabæna, Clathrocystis, and Cœlosphærium seldom give trouble unless the temperature of the water is above 70° F. Aphanizomenon is more independent of temperature. It apparently prefers a lower temperature than most of the Cyanophyceæ. In some ponds it is present throughout the entire year, even when the surface is frozen. On one occasion it grew under the ice in Laurel Lake, Fitzwilliam, N. H., and became frozen into the ice to such an extent that the ice-cutters were alarmed at the green color. In Lake Cochituate, Aphanizomenon reaches its greatest growth in the autumn. This accounts for the maximum of the curve of Cyanophyceæ in Fig. 14 occurring in October instead of in August or September.

Schizomycetes and Fungi.—These forms have no well-marked periods of seasonal distribution. They are liable to be found at any season. Mold hyphæ are frequently found at the bottom of lakes during the summer, and at the surface under the ice in winter. Crenothrix may be found in the stagnant water at the bottom of a deep lake during the summer, and at all depths in the autumn after the overturning of the lower layers of water. Crenothrix has been observed during the summer in swamps in company with Anabæna and other Cyanophyceæ.

Protozoa.—The seasonal distribution of the Protozoa, taken as an entire group, is extremely variable and differs considerably in different ponds. No curve can be drawn that will represent all cases. In Lake Cochituate the curve has a major maximum in the spring, a minor maximum in the autumn, with the summer minimum lower than that in the winter. In Mystic Lake the curve has but one maximum,—in the summer. These differences are due to the fact that the group of Protozoa is a broad one, and includes organisms that differ widely in their mode of life.

The Rhizopoda are found at all seasons of the year, but they are most numerous in the autumn after the period of summer stagnation. These organisms live upon the ooze on the bottom and sides of ponds and upon twigs and aquatic plants. There they are found most abundantly in the summer. The vertical currents of the autumnal circulation scatter them through the water and cause the maximum number of floating forms to be observed during October and November. There is a minor maximum during the period of spring circulation. Some plankton forms, such as Actinophrys, are most abundant in summer.

Of the Flagellata, Euglena, Raphidomonas and Phacus are most abundant from June to September: Trachelomonas is found at all seasons, but is most common in the fall after the period of summer stagnation: Mallomonas is found from April to October, but is usually most abundant in the autumn: Cryptomonas occurs in some ponds only in the late fall and winter: Synura and Dinobryon are generally most numerous in the spring and autumn, but heavy growths have been observed at all seasons: Uroglena seems to prefer cold weather, but vigorous growths have been noted in June.

The Dino-flagellata, Glenodinium and Peridinium, are usually most abundant during warm weather, but they are liable to occur at any season. Ceratium seldom appears before July, and it usually disappears before cold weather.

Of the Infusoria, most of the ciliated forms prefer warm water: Codonella and Tintinnus occur after periods of stagnation: Vorticella and Epistylis are distinctly summer organisms: and Bursaria and Stentor are also found in summer.

Acinteta is most abundant during warm weather.

The Protozoa that attain their greatest development in summer are those forms that are closely allied to the vegetable kingdom; namely, the Dino-flagellata and some of the Flagellata that are rich in chlorophyll. A few genera that occur most abundantly in the spring and fall have a brownish-green color like that of the diatoms, which also have spring and fall maxima. The Ciliata that live upon decaying organic matter are attuned to a comparatively high temperature,—about 75° F. This has been demonstrated by experiment, and it corresponds with the time of their observed maximum. Those Protozoa that exhibit a strictly animal mode of nutrition are most abundant at those seasons when there is plenty of food-material in the shape of minute organisms or finely divided particles of organic matter. This partially explains why growths are sometimes present in the winter when bacteria are numerous, or after periods of stagnation when particles of organic matter from the bottom have been scattered through the water.

Rotifera.—Rotifera are found at all seasons of the year, but are most numerous between June and November. In many ponds the maximum occurs in the autumn. Some genera are perennial, others are periodic in their occurrence. Anuræa and Polyarthra are found throughout the year, but their numbers rise and fall at intervals corresponding to the hatching season. Conochilus is often abundant in June, Asplanchna in July and August, and Synchæta in August and September. The littoral Rotifera are most abundant during the summer.

The Rotifera feed upon the smaller microscopic organisms, and their seasonal distribution is largely influenced by the amount of this food-supply. The reactions of the Rotifera to light, temperature, etc., are not well known.

Crustacea.*—The number of Crustacea present at differ-

---

* For a full discussion of the seasonal distribution of the Crustacea the reader is referred to Dr. Birge's studies of the Crustacea of Lake Mendota.

ent seasons varies greatly in different bodies of water. It is influenced largely by the genera that are present. Different genera vary considerably in their seasonal distribution. Some are found at all seasons, while others occur only at certain times. The perennial forms may have several maxima during the year, corresponding to the hatching of different broods. As a rule Crustacea are most numerous in the spring, but minor maxima may occur during the summer and autumn and rarely in the winter.

Temperature, food-supply, and competition are said to be the chief factors that influence the seasonal distribution of the Crustacea.

CHAPTER VIII.

HORIZONTAL AND VERTICAL DISTRIBUTION OF MICROSCOPIC ORGANISMS.

THE plants and animals that inhabit lakes and ponds may be classified according to their habitat, but it is sufficient here to consider them either as *littoral* or *limnetic*.

The *littoral* organisms may be said to include all those forms that are attached to the shore or to plants growing on the shore, besides a host of others which, though free-swimming, are almost invariably associated with the attached forms.

The *limnetic* or *pelagic* organisms are those that make their home in the open water. They float or swim freely and are drifted about by every current. Collectively they make up the greater part of the plankton. They include almost all the troublesome odor-producing organisms in water-supplies. In the open water, however, one often finds some of the littoral forms that have been detached from the shore and scattered through the water by the currents, or that are parasitically attached to some of the limnetic forms. Then there are organisms that may be said to be *facultative limnetic forms*, that is, they are sedentary or free-swimming at will. The true limnetic forms, however, are the most important in water-supplies, and their horizontal and vertical distributions are now to be considered.

Horizontal Distribution.—The horizontal distribution of the limnetic organisms is usually quite uniform within any limited area, but through the entire body of a lake the number of organisms may show considerable variation. This is quite noticeable in long, narrow reservoirs that have streams entering at one end and discharging at the other. In such reservoirs the organisms are generally most numerous at the lower end. If, however, the water in the influx stream contains many organisms the numbers may be higher at the upper end, diminishing gradually as the water of the stream becomes mixed with that of the reservoir. Sometimes the mixing takes place slowly and the influent water passes as a current far into the reservoir. This tends to distribute the organisms in streaks. In lakes with uneven margins the horizontal distribution may vary greatly, and the number of organisms found in coves may be quite different from the number found in the open water. The horizontal distribution of diatoms is influenced to some extent by the depth of the lake. There is in Massachusetts a lake covering about 250 acres. Near one side of it there is a deep hole, that has an area of about five acres, where the stagnation-phenomena are very pronounced. When the growths of diatoms occur in the spring and fall the numbers are very much higher in the vicinity of this deep hole than elsewhere in the lake.

Areas of shallow flowage exert a marked effect on the horizontal distribution of the microscopic organisms.

The wind also has a great influence, and in many bodies of water it is the controlling influence. The organisms, particularly the Cyanophyceæ, are driven in the direction of the wind and accumulate towards the lee shore. It is possible that horizontal thermal convection currents may influence the distribution to some extent.

Vertical Distribution.—The laws that govern the vertical distribution of the microscopic organisms are more complicated than those which govern their horizontal distribution. The latter affect the organisms mechanically; the former, vitally. While their specific gravity and the vertical currents produced mechanically or thermally play an important part, the amount of food-material and dissolved oxygen and the amount of heat and light influence the very life of the organisms.

In a lake of the second order the determining factors vary at different depths and at different seasons. In the summer, for example, the conditions above the thermocline * are very different from those below it. Near the surface the water is warm, the light is strong, oxygen is very abundant, and there are vertical currents. Near the bottom the water is cold, the light is weak, the oxygen may be exhausted, and the water is perfectly quiet. With these conditions chlorophyll-bearing organisms naturally thrive best above the thermocline. They seldom develop below it.

It has been shown by experiment that the development of diatoms is greatest near the surface and that it decreases downwards as the light decreases. In nature, however, it cannot be expected that the number of diatoms in the different layers of water will follow this law closely, because the diatoms are heavy and constantly tend to sink and because the water above the thermocline is more or less stirred up. One would expect rather to find a uniform vertical distribution above the thermocline, and below it a rapid decrease in the number of organisms. Such a distribution is common. The following instances of the vertical distribution of Asterionella and Tabellaria in Lake Cochituate may be cited in illustra-

---

* See page 62.

tion: in both instances the thermocline was located between 20 and 30 ft.

VERTICAL DISTRIBUTION OF ASTERIONELLA AND TABELLARIA IN LAKE COCHITUATE.

NUMBERS PER C.C.

Depth in Feet.	Asterionella. May 7, 1891.	Tabellaria. May 24, 1890.
Surface	3752	1886
10 ft.	3736	1448
20 "	3716	1396
25 "	—	484
30 "	1784	298
40 "	456	—
50 "	536	—
60 "	178	96

This manner of distribution is most common during periods of rapid development, when a gentle breeze is stirring. In very quiet weather and during periods of declining growth diatoms sink rapidly, and at such times they may be found most numerous at the thermocline or at the bottom. During periods of complete vertical circulation the vertical distribution may be quite uniform from top to bottom. The diatoms found at the bottom of a deep lake are usually less vigorous than those near the surface.

The Chlorophyceæ and Cyanophyceæ are much lighter in weight than the diatoms, and some of them contain oil-globules and bubbles of gas. The forces tending to keep them near the surface are greater therefore than in the case of the diatoms. These forms are seldom found below the thermocline, and even above the thermocline they show considerable variations at different depths. The Cyanophyceæ especially collect near the surface. In quiet waters they often form unsightly and ill-smelling scums. Occasional exceptions to the general rule are observed. Microcystis, for example, is usually more abundant in Lake Cochituate just below the thermocline than it is at the surface. On July 31, 1895, the numbers of standard units of Microcystis at different depths were as follows: Surface, 94; 30 ft., 342; 60 ft., 140.

It is interesting to notice that a sudden wind affects the vertical distribution of the Cyanophyceæ and the Diatomaceæ in opposite ways. It tends to decrease the number of blue-green algæ at the surface by preventing the formation of scums, while it increases the number of diatoms by preventing them from sinking.

The Protozoa, as a class, seek the upper strata of water. Euglena sometimes form a scum upon the surface. Uroglena, Synura, etc., are often most numerous in winter just beneath the ice. The Dino-flagellata are distinctly surface forms. Some of the Protozoa seem to avoid direct sunlight and keep away from the upper surface of the water, though they may be very abundant at a depth of one or two feet. The Ciliata and those Protozoa that have a distinctly animal mode of nutrition are more irregularly distributed through the vertical. The Rhizopoda are most abundant near the bottom.

At times some of the Protozoa are more numerous at the thermocline than elsewhere in the vertical. An interesting illustration of this occurred in Lake Cochituate in the summer of 1896. Mallomonas are not ordinarily abundant in this lake, but on June 24 they suddenly appeared just below the thermocline. At the mid-depth (30 ft.) there were 116 per c.c., at the bottom there were 42 per c.c., but at the surface there were none. They developed rapidly, and on August 4 there were 3640 at the mid-depth. The growth continued until September, and during this time the largest number observed at the bottom was 276 per c.c., while above the thermocline scarcely an individual was found. On July 17 the vertical distribution was as follows:

VERTICAL DISTRIBUTION OF MALLOMONAS IN LAKE COCHITUATE, JULY 17, 1896.

Depth.	Number per c.c.	Temperature Fahr.
Surface	0	77.3°
10 ft.	0	75.2
15 "	2	62.0
20 ft.	1454	47.7
25 "	794	43.7
30 "	548	43.2
40 "	112	42.5
50 "	88	41.4
60 "	64	40.8

Synura and other Protozoa have sometimes shown a similar vertical distribution. Whether this concentration at the thermocline is due to food-material, to light, or to temperature is not definitely known. Mallomonas are motile and are known to be positively heliotropic. In the winter they are often numerous under the ice. It is possible that they have a low temperature attunement, and that in the instance above cited they collected as near the surface as their temperature attunement would permit. This would accord with the fact that they are most numerous in the spring and fall.

Rotifera and Crustacea are most numerous above the thermocline, and as a rule they are concentrated in the upper strata of water. During the winter they are sometimes abundant at the bottom. Different genera react differently to light, heat, etc., and therefore the vertical distribution of these organisms is somewhat complicated. Some of them show a slight daily migration towards the surface at night, and away from the surface in the daytime.

The Schizomycetes are usually more abundant at the bottom of a pond than at the surface. Mold hyphæ are often numerous in winter just under the surface of the ice.

In spite of the tendencies of the organisms to choose their favorite habitat in a body of water, the mechanical effects of winds, currents, gravity, etc., are so great that in most ponds and reservoirs used for water-supply (except very deep ones) the average number of organisms of all kinds through the year does not vary much at different depths. This is illustrated by the following table:

TABLE SHOWING THE RELATIVE NUMBER * OF MICROSCOPIC ORGANISMS OF ALL KINDS AT THE SURFACE, MID-DEPTH, AND BOTTOM OF THE RESERVOIRS OF THE BOSTON WATER WORKS.

Locality.	Depth.	1890.	1891.	1892.	1893.	1894.	1895.	1896.
Lake Cochituate	Surface 30 ft. 60 ft.	454 304 357	736 569 650	523 528 626	389 336 316	416 365 309	355 373 353	507 657 544
Basin 2	Surface 13 ft. 25 ft.	68 80 64	322 273 268	268 256 229	116 98 98	45 49 33	61 56 47	87 120 78
Basin 3	Surface 15 ft. 30 ft.	152 182 131	277 267 323	514 523 481	381 303 311	289 194 179	621 543 485	524 467 498
Basin 4	Surface 20 ft. 40 ft.	50 38 25	129 95 83	269 268 235	112 84 66	28 20 20	57 35 25	94 108 106
Basin 6	Surface 25 ft. 50 ft.				87 52 72	105 58 53	189 118 104

* For the years 1890 to 1893 the results were given in "Number of Organisms per c.c." Since Jan. 1, 1893, the results have been given in Number of Standard Units per c.c. (One standard unit equals 400 square microns.)

The vertical distribution varies at different seasons, as the following table illustrates:

TABLE SHOWING THE RELATIVE NUMBER OF ORGANISMS (STANDARD UNITS) PER C.C. AT THE SURFACE, MID-DEPTH, AND BOTTOM OF THE RESERVOIRS OF THE BOSTON WATER WORKS DURING 1895.

Locality.	Depth.	January.	February.	March.	April.	May.	June.	July.	August.	September.	October.	November.	December.	Mean.
Lake Cochituate	Surface	255	34	10	97	188	437	480	248	137	450	1159	762	355
	30 ft.	407	21	23	109	149	188	539	329	193	400	1199	921	373
	60 ft.	422	232	55	101	133	188	503	290	53	252	1198	808	353
Basin 2	Surface	6	8	6	49	56	109	163	152	82	72	15	18	61
	13 ft.	6	7	18	25	59	76	195	108	93	53	14	17	56
	25 ft.	4	7	17	22	47	63	160	88	74	49	22	9	47
Basin 3	Surface	13	3	14	62	375	787	1197	1675	1778	1227	266	53	621
	15 ft.	18	1	14	46	260	768	1072	1134	1813	1161	253	34	543
	30 ft.	47	4	13	57	235	597	633	1146	1487	1342	222	37	465
Basin 4	Surface	78	74	10	27	79	76	123	75	30	45	40	22	57
	20 ft.	18	19	5	15	37	47	43	78	29	55	37	19	35
	40 ft.	13	19	18	12	21	41	48	38	7	33	21	26	25
Basin 6	Surface	41	50	36	64	91	193	203	91	243	186	41	13	105
	25 ft.	28	10	4	57	42	61	46	85	130	190	56	9	58
	50 ft.	4	5	21	76	51	39	18	47	83	214	60	16	53

A further analysis of the results at Lake Cochituate shows the vertical distribution of the different classes of organisms to be as follows:

RELATIVE NUMBER OF ORGANISMS (STANDARD UNITS) PER C.C. AT THE SURFACE AND BOTTOM OF LAKE COCHITUATE.

AVERAGE FOR THE YEAR 1895.

	Diatomaceæ.	Chlorophyceæ.	Cyanophyceæ.	Protozoa.	Rotifera.	Miscellaneous.	Total.
Surface	144	79	108	17	3	4	355
Bottom	160 *	16	67	10	1	99 †	353

* If the dead and empty cells were excluded this figure would be much lower.
† Chiefly Crenothrix.

CHAPTER IX.

ODORS IN WATER-SUPPLIES.

The senses of taste and odor are distinct, but they are closely related to each other. There are some substances, like salt, that have a taste but no odor, and there are other substances, like vanilla, that have a strong odor but no taste. It is believed that the sense of taste is quite limited and that many so-called tastes are really odors, the gas or vapor given off by the substance tasted reaching the nose not only through the nostrils but through the posterior nares. Thus an odor tasted is often stronger than an odor smelled.

Chemically pure water is free from both taste and odor. Water containing certain substances in solution, as sugar, salt, etc., may have a decided taste but no odor. Such taste-producing substances are met with in mineral waters or in brackish or chalybeate waters, but as a rule they are not offensive and they seldom affect large bodies of water. Most of the bad tastes observed in drinking-water are due not to inorganic but to organic substances either in solution or in suspension. Such substances almost invariably produce odors as well as tastes. The subject may be pursued therefore from the standpoint of odor, though in many instances the best way to observe the odor of the water is to taste it.

Water taken directly from the ground and used immediately is usually odorless. In certain sections of the country it has a sulphurous odor. If it is contaminated or drawn from a swampy region it may be somewhat moldy or unpleasant.

Almost all surface-waters have some odor. Many times it is too faint to be noticed by the ordinary consumer, though it can be detected by one whose sense of smell is carefully trained. On the other hand, the water in a pond may have so strong an odor that it is offensive several hundred feet away. Between these two extremes one meets with odors that vary in intensity and in character, and that are often the source of much annoyance and complaint.

It is difficult to classify the odors of surface-waters on a satisfactory basis, but they fall into three general groups: 1. Odors caused by organic matter other than living organisms. 2. Odors caused by the decomposition of organic matter. 3. Odors caused by living organisms.

1. The odors caused by organic matter other than living organisms may be included under the general term *vegetable*. They vary in character in different waters and at different seasons. It is difficult to find terms that will describe them exactly. It is seldom that two observers will agree as to the most appropriate descriptive adjective. To one person the odor of a water may be *straw-like*, to another *swamp-like*, to another *peaty*. This is due to the fact that the sense of smell in man is not well cultivated. In practice, therefore, it has become customary to use the general term *vegetable* instead of the terms *straw-like, swamp-like, marshy, peaty, sweetish*, etc. The intensity of an odor may be indicated by using the prefixes *very faint, faint, distinct, decided, very strong*. In a general way the following values may be assigned to these terms, which are applied not only to the vegetable odors but to the odors of the other groups as

well. A *very faint* odor is one that would not be detected by the ordinary consumer. A *faint* odor is one that would be detected if attention were called to it but that otherwise would not attract notice. A *distinct* odor is one that would be readily detected and that might cause the water to be looked upon with disfavor. A *decided* odor is one that would force itself upon the attention and that might make the water unpalatable. The term *"very strong"* is reserved for those odors that make a water unfit for drinking. This term is seldom used. The reader will understand that these definitions are far from exact, and that the intensity of odors varying in character cannot be well compared. A *faint fishy* odor, for example, might often attract more attention than a *distinct vegetable* odor. Heating a water usually intensifies its odor.* A water that has a *faint* odor when cold may have a *distinct* odor when hot.

Most of the *vegetable* odors are caused by vegetable matter in solution. Brown-colored waters invariably have a *sweetish-vegetable* odor, and the intensity of the odor varies almost directly with the depth of the color. Both color and odor are due to the presence of certain glucosides, of which tannin is an example, extracted from leaves, grasses, mosses, etc. In addition to the odor, these substances have a slight astringent taste. Colorless waters containing organic matter of other origin may have *vegetable* odors, but they are usually less *sweetish* and more *straw-like* or *peaty*. Akin to the *vegetable* odors are the *earthy* odors caused by finely divided

---

* In the laboratory, the "cold odor" is observed by shaking a partly filled bottle of the water and immediately removing the stopper and applying the nose. The "hot odor" is obtained by heating a portion of the water in a tall beaker covered with a watch-glass to a point just short of boiling. When sufficiently cool the cover is slipped aside and the observation made.

particles of organic matter, clay, etc. The two odors are often associated in the same sample.

2. Odors produced by the decomposition of organic matter in water are not uncommon. They are described, somewhat imperfectly, by such terms as *moldy, musty, unpleasant, disagreeable, offensive*. An *unpleasant* odor is produced when the vegetable matter in water begins to decay. It may be said to represent the first stages of decomposition. As decomposition progresses the *unpleasant* odors become *disagreeable*, and then *offensive*. It is seldom that the decomposition of vegetable matter in water produces odors worse than "*decidedly unpleasant.*" The *disagreeable* odors usually can be traced to decaying animal matter, and, as a rule, *offensive* odors are observed only in sewage or in grossly polluted water. The terms *moldy* and *musty* are more specific than the terms *unpleasant, disagreeable*, and *offensive*, but they are difficult to define. They are quite similar in character; but the *musty* odor is more intense and is usually applied only to sewage-polluted water. The *moldy* odor suggests a damp cellar, or perhaps a decaying tree-trunk in a forest. The bacteriologist will recognize this odor as similar to that given off by certain bacteria growing on nutrient gelatine.

The odors of decomposition naturally are associated with the odors of the other groups, and one often finds it convenient to use such expressions as "*distinctly vegetable and faintly moldy,*" or "*decidedly fishy and unpleasant.*"

3. The odors of drinking-water due to the presence of living organisms are the most important because of their common occurrence, because of their offensive nature, and because they affect large bodies of water. It is only within recent years that these odors have been well understood, and even now there is much to be learned about the chemical

nature of the odoriferous substances and their relation to the life of the organisms. At one time it was supposed that it was only by decay that the organisms became offensive. It is now a well-established fact that many living organisms have an odor that is natural to them and that is peculiar to them, just as a fresh rose or an onion has a natural and peculiar odor. It has been found, also, that in most cases,—and it may be true in all cases,—the odor is produced by compounds analogous to the essential oils. In some cases the oily compounds have been isolated by extraction with ether or gasoline. Odors due to these oils have been called "odors of growth" because the oils are produced during the growth of the organisms. The oil-globules may be seen in many genera if they are examined with a sufficiently high power. They are usually most numerous in the mature forms and are often particularly abundant just before sporulation or encystment. The production of the oil represents a storing-up of energy. The odors have been called "odors of disintegration," because they are most noticeable when the breaking up of the organism causes the oil-globules to be scattered through the water. It is sufficient, however, to call them the "natural odors" of the organisms, to distinguish them from the very different odors produced by their decomposition.

It was stated in Chapter IV that the microscopic organisms are not found in ground-waters (except when stored in open reservoirs) nor in streams in sufficient abundance to cause trouble. It is in the quiescent waters of ponds and lakes and reservoirs that they develop luxuriantly, and it is to the reservoir that one should look first when investigating the cause of an odor in a public water-supply.

The littoral organisms found on the sides of reservoirs

include the flowering aquatic plants, the Characeæ, the filamentous algæ, etc., of the vegetable kingdom and the fresh-water sponge, Bryozoa, etc., of the animal kingdom. The effect which they exert on the odor of a water is difficult to determine because they are seldom found in a reservoir where the floating microscopic organisms are absent. In many cases where a peculiar odor of a water has been charged to some of these littoral forms, subsequent investigation has made it probable that the odor was really caused by limnetic organisms that had been overlooked in the first instance.

Speaking generally it may be said that in reservoirs that are large and deep the organisms attached to the shores produce little or no effect on the odor of the water; and that in small shallow reservoirs where the aquatic vegetation is thick they do not impart any characteristic "natural" odor, but they may produce a sort of vegetable taste and a disagreeable odor due to decomposition.

Some of the littoral aquatic plants, such as Myriophyllum and a number of the filamentous algæ, possess a natural odor that is strongly "vegetable" and, at times, almost fishy; but the odor is obtained only when the plants are crushed or when fragments are broken off and scattered through the water. Under ordinary conditions of growth in a reservoir this does not happen and therefore no odor is imparted to the water except through decomposition.

There are on record some apparent exceptions to the rule that the attached growths cause no odor. Hyatt* described a growth of Meridion circulare at the headwaters of the Croton River, in 1881, that was supposed to have affected the entire supply of New York City: Rafter has connected odors with

---

* References to this and similar illustrations may be found in the bibliography in the appendix.

Hydrodictyon utriculatum and other Chlorophyceæ: Forbes investigated a water-supply where a growth of Chara was thought to be the cause of a bad odor; and Weston has stated that serious trouble was caused in Henderson, N. C., by an extensive growth of Cristatella.* All of these cases where odors in water-supplies have been attributed to certain limnetic organisms lack corroboration.

The author once examined a reservoir where a mass of Melosira varians several feet thick covered the slopes to a considerable depth. A severe storm tore away the fragile filaments, and masses of Melosira passed into the distribution-pipes and caused a noticeable vegetable and oily odor in the water.

In connection with the relation of the littoral organisms to odors in water-supplies some reference should be made to the "cucumber taste" that has been a frequent cause of complaint against the Boston water-supply. In 1881 the trouble was very severe. The water had a decided odor of cucumbers, which was intensified at times to a "fish-oil" odor. Heating made the odor very strong and offensive. A noted expert made an examination and concluded that the seat of the trouble was in Farm Pond,—one of the sources of supply. This pond was so situated that all the water of the Sudbury system passed through it on its way to the city. Chemical analysis of the water and microscopical examination of the mud failed to reveal the cause of the odor. It was found, however, that fragments of fresh-water sponge (Spongilla fluviatalis) were constantly collecting on the screens and that these had the "cucumber odor." It was decided therefore that the fresh-water sponge was the cause of the odor. The conclusion was quite generally accepted and the report has been quoted extensively.

---

* The organism observed was probably Pectinatella and not Cristatella.—AUTHOR.

At that time some water experts disagreed with this opinion. They claimed that the amount of sponge found in the pond was not sufficient to produce the odor. In the light of modern microscopical examinations we are coming to believe that the dissenters were right and that the fresh-water sponge was not the cause of the cucumber odor. The author has taken masses of Spongilla and allowed them to rot in a small quantity of water till the odor was unbearable. This water was then diluted with distilled water to see how large a mass of water the decayed sponge would affect. It was found that with a dilution of 1 to 50 000 there was no perceptible odor. If this is true it would take a mass of sponge several feet thick over the entire bottom of Farm Pond to produce an odor as intense as that observed in 1881. Morever the odor produced by the sponge is not the "cucumber odor," although it is similar to it.

There is good reason to believe that the cucumber odor observed in 1881 was due to Synura. One need not dispute the observation that the sponge that collected on the Farm Pond screens had the cucumber odor, for no doubt the sponge was covered with Synura, as it is often covered with other organisms. It is not surprising, either, that the Synura should have been overlooked in the water, because the organism disintegrates readily and a comparatively small number of colonies is able to produce a considerable odor. The times of the occurrence of the odor—namely, in the spring and autumn—are worth noting, as they correspond with the seasons when Synura grows best and when it is most commonly found.

In February, 1892, the cucumber taste again appeared in the Boston water. This time it was definitely traced to Synura that was growing in the water just under the ice in

Lake Cochituate. Since then it has reapppeared at intervals in other parts of the supply,—notably in Basin 3 and Basin 6. It has been found that 5 or 10 colonies per c.c. are sufficient to cause a perceptible odor.

The floating microscopic organisms, or the plankton, are responsible for most of those peculiar nauseating odors that are the cause of complaint in so many public water-supplies. In most, if not in all, cases the odor is due to the presence of an oily substance elaborated by the organisms during their growth. This has been proved by long-continued observations and experiments, during the course of which the following facts have been noted:

The odors referred to vary in character. They are difficult to describe, but they can be readily identified. Particular odors are associated with particular organisms. If an organism is present in sufficient numbers its particular odor will be observed; if it is not present in sufficient numbers its odor will not be observed. Further, the intensity of the odor varies with the number* of organisms present. If water that contains an organism which has a natural odor is filtered through paper, the odor of the filtered water † will be much fainter than before, and the filter-paper on which the organisms remain will have a strong odor. If the organisms are concentrated by the Sedgwick-Rafter method, the concentrate will have a decided taste and odor. If these organisms are placed in distilled water, the water will acquire the odor of the original water. Thus, the relation between particular odors and particular organisms has been well established. Indeed, in the absence of a microscopical examination, experienced

---

* There are some exceptions to this.

† In some cases the odoriferous substances from the organisms pass through the filter, and the disintegration of the organisms gives the filtered water an increased odor over the unfiltered water.

observers are often able to tell the nature of the organisms present by a simple observation of the odor.

That the odors are not due to the decomposition of the organism is proved by the character of the odors themselves and by the fact that they are not accompanied necessarily by large numbers of bacteria or by the presence of free ammonia or nitrites. This is supported by the fact that, when the organisms do decay, the bacteria increase in number and the odor of the water changes in character.

The natural odor is given off by some substance inside of the organism, and when this substance becomes liberated the odor is more easily detected. The odor is intensified by heating, by mechanical agitation, and by change in the density of the water containing the organisms. Many of the odor-producing organisms are very delicate. Heating breaks them up and drives off the odoriferous substances. The flow of water through the pipes of a distribution system is sufficient to cause the disintegration of many forms, and it is a matter of common observation that in such cases the odor of a water at the service-taps is more pronounced than at the reservoir. If the density of a water is increased by adding to it some substance, such as salt, the organisms may become distorted if not actually broken up. This causes an intensification of their odor. Increased pressure also leads to the same result.

The natural odor of the organisms is due to some oily substance analogous to those substances found in higher plants and animals, and that give the odor to the peppermint and the herring. The fact was noted long ago that the addition of salt to water that was affected with certain odors developed an oily flavor. Many of the odors caused by organisms are of a marked oily nature. The oil-globules in these organisms may be observed with the microscope. The number of oil-globules varies according to the age and condition of the organisms, and the intensity of the odor varies with the number of oil-globules present. Finally, the oily substances have been extracted from the organisms and it has been found that they possess the same odor as that observed in the water containing them.

A series of experiments was made at one time to show that the amount of oil present in the organisms was sufficient to account for the odors observed in drinking-water. Some of the familiar essential oils, such as oil of peppermint, oil of clove, cod-liver oil, etc., were diluted with distilled water, and the amount of dilution at which the odor became unrecognizable was noted. The oil of peppermint was recognized when diluted 1 : 50 000 000; the oil of clove, 1 : 8 000 000; cod-liver oil, 1 : 1 000 000; etc. The odor of kerosene oil could not be detected when diluted 1 : 800 000. The amount of oil present in water containing a known number of organisms was estimated for comparison. It was found that in water containing 100 colonies of Synura per c.c. the dilution of the Synura oil was 1 : 25 000 000; and that in a water with 50 000 Asterionella per c.c. the dilution was only 1 : 2 000 000. Thus, the production of the odor by the oil is quite within the range of possibility. An interesting fact brought out by the experiments was that the odor of the oils varied with different degrees of dilution not only in intensity but in character.* This variation of the character of the odor with its intensity is important to notice, as it accounts for the different descriptions of the same odor in a water-supply at different times and by different people.

---

* On one occasion seven people out of ten who were asked to observe the odor of very highly diluted kerosene oil declared that it smelled like "perfumery."

The nature of the odoriferous oils or oily substances is not well known. Calkins, who isolated the odoriferous principle of Uroglena with gasoline and ether, describes it as being similar to the essential oils. It was non-volatile at the temperature of boiling water. Jackson and Ellms extracted a similar substance from Anabæna with gasoline. On standing it oxidized and became resinous. It contained needle-like crystals. Experiments by the author have shown that the oils of Asterionella and Mallomonas are quite similar in character.

Most, if not all, of the organisms produce oil during their growth to a greater or less degree. In many cases it is quite odorless. Water is often without odor even when large numbers of organisms are present. This is either because the organisms have not produced oil, or because the oil is odorless. Sometimes water rich in organisms will have an oily flavor with no distinctive odor. This is true in the case of some species of Melosira. Many organisms impart a vegetable and oily taste, without a distinctive odor. This is true of Synedra pulchella and Stephanodiscus. There are, moreover, microscopic organisms that produce oils that have a distinctive odor, but that occur in drinking-water in such small numbers that the odor is not detected. The organisms that have a distinctive odor and that are found in large numbers are comparatively few. Not more than twenty-five have been recorded and only about half a dozen have given serious trouble. More extended observations may lengthen this list.

The distinctive odors produced by these organisms may be grouped around three general terms,—*aromatic, grassy,* and *fishy,*—and for convenience they may be tabulated as follows:

Group.	Organism.	Natural Odor.
Aromatic Odor.	Diatomaceæ	
	Asterionella	Aromatic—geranium—fishy.
	Cyclotella	Faintly aromatic.
	Diatoma	" "
	Meridion	Aromatic.
	Tabellaria	"
	Protozoa	
	Cryptomonas	Candied violets.
	Mallomonas	Aromatic—violets—fishy.
Grassy Odor.	Cyanophyceæ	
	Anabæna	Grassy and moldy—green-corn—nasturtiums, etc.
	Rivularia	Grassy and moldy.
	Clathrocystis	Sweet, grassy.
	Cœlosphærium	" "
	Aphanizomenon	Grassy.
Fishy Odor.	Chlorophyceæ	
	Volvox	Fishy.
	Eudorina	Faintly fishy.
	Pandorina	" "
	Dictyosphærium	" "
	Protozoa	
	Uroglena	Fishy and oily.
	Synura	Ripe cucumbers—bitter and spicy taste.
	Dinobryon	Fishy, like rockweed.
	Bursaria	Irish moss—salt marsh—fishy.
	Peridinium	Fishy, like clam-shells.
	Glenodinium	Fishy.

The aromatic odors are due chiefly to the Diatomaceæ. The strongest odor is that produced by Asterionella. The character of this odor changes with its intensity. When few organisms are present the water may have an undefinable *aromatic* odor; as they increase the odor resembles that of a *rose geranium;* when they are very abundant the odor becomes *fishy* and *nauseating*. The other diatoms given in the table produce the aromatic odor only when present in very large numbers. There are two Protozoa that have an aromatic odor. The odor of Cryptomonas is *sweetish* and resembles that of the *violet*. The odor of Mallomonas is similar to that of Cryptomonas, but when strong it becomes *fishy*.

The grassy odors are produced by the Cyanophyceæ. Anabæna is the most important organism of this class. There are several species that have slightly different odors. The *grassy* odor is usually accompanied by a *moldy* odor, which is probably due to decomposition, as this organism decays rapidly. When very strong the odor of Anabæna much resembles *raw green-corn*, or even a *nasturtium* stem. The prevailing odor, however, is *grassy*, i.e. the odor of freshly cut grass. The other blue-green algæ have odors that may be called *grassy*, but they are less distinctive than in the case of Anabæna.

The *fishy* odors are the most disagreeable of any observed in drinking-water. That produced by Uroglena is perhaps the worst. It is quite common. Water rich in Uroglena has an odor not unlike that of *cod-liver oil*. The odor of Synura is almost as bad and almost as common. It resembles that of a *ripe cucumber*. Synura also has a distinct bitter and spicy taste. It "stays in the mouth" and is most noticeable at the back part of the tongue. Glenodinium and Peridinium both produce fishy odors. The latter somewhat resembles *clam-shells*. Dinobryon has a fishy odor and suggests *sea-weed*. The odor of Bursaria is like that of *Irish moss*. It also reminds one of a *salt marsh*. With certain degrees of dilution some other Protozoa have the salt-marsh odor, reminding one of the sea. Fishy odors are said to be produced by Volvox, Eudorina, and Pandorina. These Chlorophyceæ are sometimes classed with the Protozoa, so that it may be said in a general way that the fishy odors are produced by microscopic organisms belonging to the animal kingdom.

Some of the microscopic organisms have distinctive odors of decomposition. The Cyanophyceæ when decaying give a "pig-pen" odor. Beggiatoa and some species of Chara give the odor of sulphuretted hydrogen. All the odors given off by the decomposition of microscopic organisms are offensive. They are particularly so when the organisms contain a high percentage of nitrogen. Jackson and Ellms, in an interesting study of the decomposition of Anabæna circinnalis, found that that organism contained 9.66% of nitrogen. They found that the "pig-pen" odor was due "to the breaking down of highly organized compounds of sulphur and phosphorus and to the presence of this high percentage of nitrogen. The gas given off during decomposition was found to have the following composition:

Marsh-gas	0.8%
Carbonic acid	1.5%
Oxygen	2.9%
Nitrogen	12.4%
Hydrogen	82.4%
	100.0%

The gas that remained dissolved in the water containing the Anabæna was practically all CO_2 and represented a large percentage of the total gas produced."

Besides the odors above described, water-supplies sometimes become affected with what have been called "chemical odors,"—such as those of carbolic acid, creosote, tar, etc. They can be traced usually to some pollution by manufacturing waste, though a vigorous decomposition of organic matter has been known to give an odor resembling carbolic acid. Similar odors are sometimes caused by the coating on the inside of new distribution-pipes.

The extent to which water-supplies are afflicted with odors

was well shown by the investigations of the Massachusetts State Board of Health. Out of 71 water-supplies taken from ponds and reservoirs, 45, or 63%, were found to have given trouble from bad tastes or odors, and about two thirds of these had given serious trouble. Calkins has stated that in 1404 samples from surface-water supplies in Massachusetts odors were observed as follows:

Odor.	Per Cent of Samples Affected.
No odor	20
Vegetable	26
Sweetish	7
Aromatic	6
Grassy	15
Fishy	3
Moldy	10
Disagreeable	6
Offensive	7

The intensity of these odors was not stated. Many of them probably were not strong enough to cause complaint.

It must not be inferred from this that Massachusetts is more afflicted in her surface-water supplies than other sections of the country. The same troubles are observed everywhere. It is only because the Massachusetts supplies have been more carefully studied than elsewhere that attention has been drawn to them. In a previous chapter it was stated that the microscopic organisms are widely distributed both in this country and abroad. Wherever they are found in abundance they must inevitably affect the odor of the water.

The question is often asked, "Are growths of organisms such as Asterionella, Synura, etc., injurious to health?" This cannot be answered authoritatively, but from the data at hand it is believed that such organisms are not injurious,— certainly not to persons in good health. The actual amount of solid matter contained in the organisms is much smaller than might be supposed. For example, it has been calculated that the weight of one Asterionella is .000000004 gram. A growth of 100 000 Asterionella per c.c. would render a water unfit to drink because of its odor, yet a tumblerful of such water would contain but eight milligrams of solid matter, and only one half of this would be organic matter. It is almost inconceivable that such a small amount of organic matter could cause trouble unless some poisonous principle were present, and so far as is known no such substance has been found. The alleged cases of poisonous algæ rest upon too uncertain evidence to be received as facts.

Nevertheless there is some reason to believe that people accustomed to drinking-water free from organisms may be subjected to temporary intestinal disorders when they begin to drink water rich in microscopic organisms,—just as people are affected by changing from a hard to a soft water and *vice versa*. It is possible that with young children and invalids such disorders may be more common than has been supposed.

The subject of the removal from drinking-water of the odors produced by microscopic organisms is not treated in this work because the results thus far obtained do not warrant publication. Ordinary filtration may not always prove successful, because the odor-producing substances are sometimes capable of passing unchanged through sand layers of considerable thickness. Aeration after filtration is beneficial to a limited extent.

CHAPTER X.

STORAGE OF SURFACE-WATER.

To obtain a permanently safe and satisfactory surface-water supply without filtration the rainfall must be collected quickly from a clean watershed and stored in a clean reservoir.

A clean watershed may be defined as one upon which there are no sources of pollution and no accumulations of decomposing organic matter. The subject of pollution is of paramount importance, but it will not be emphasized here as its discussion leads into bacteriology rather than into microscopy. No watershed can be free from organic matter, and this must eventually decompose. The grass dies, the leaves fall, and a thin layer of decay is spread over the surface of the ground. This is repeated each season. Normally this organic matter disappears by rapid oxidation, and if the ground is sloping the rain that falls upon it runs off rapidly and absorbs comparatively little organic matter. If, however, the decaying vegetation has accumulated in thick layers, if the ground is level and becomes saturated or covered with water, decomposition takes place under different conditions, and the water may become highly charged with organic matter and the products of decay.

The effect of swamp areas upon the color of water has been referred to. Water from a clean watershed seldom has a color higher than .30 of the Platinum Scale. The amount of color above this figure can be generally traced to swampy land. The color of the stagnant water of swamps is sometimes very high,—often 3.00 and sometimes as high as 5.00 or 7.00 on the "Platinum Scale." From this it is easy to see that even a comparatively small percentage of swamp-land upon a watershed may have an important effect upon the color of the combined yield.

A highly colored water means a water rich in organic matter. If the color is much above .50 the water has an unsightly appearance, a distinct vegetable odor, and a sweetish and somewhat astringent taste. But the presence of organic matter is objectionable for another reason. It helps to furnish food-material for the microscopic organisms, and these may render the water very disagreeable. Swamps are breeding-places for many of the organisms that cause trouble in water-supplies, and numerous instances might be cited where organisms have developed in a swamp and have been washed down into a storage-reservoir, rendering the water there almost unfit for use.

Cedar Swamp, at the head of the Sudbury River of the Boston water-supply, furnishes an example of this. During August, 1892, Anabæna developed abundantly in a small pond in the middle of this swamp. At one time there were 8400 filaments (about 50 000 standard units) per c.c. A heavy rain washed the Anabæna down-stream, and on August 15 there were 2064 filaments per c.c. at the upper end of Basin 2. On August 17 the water entering the basin contained but 600 filaments, and a week later it contained none. The Anabæna were washed down-stream in a sort of wave. Basin 2 is a long, narrow basin. The wave of Anabæna passed through the basin, down the aqueduct, through the Chestnut Hill Reservoir, and into the service-pipes. On August 22 Anabæna were first observed at the gate-house at the lower end of Basin 2, where there were 647 filaments per c.c., and on the following day they appeared at the terminal chamber of the conduit at Chestnut Hill Reservoir, where there were 326 filaments per c.c. In another week they became disseminated through this reservoir and were found in the service-pipes. As the water from Basin 2 passed towards the city it became mixed with the water from other sources, so that by the time it reached the consumers the Anabæna were not sufficiently abundant to cause complaint. After the first wave of Anabæna had passed through Basin 2 the organisms began to increase throughout the basin, and the growth continued for several weeks. It was evident that the water from the swamp carried down not only the Anabæna themselves, but enough food-material to support their growth in the basin.

Instances are still more common where organisms from swamps have seeded storage-reservoirs. Entering the reservoir in comparatively small numbers, the organisms frequently find in the quiet water conditions favorable to their growth. Growths of some of the Flagellata may be traced directly to seeding from swamps. The draining of swamps makes a vast improvement in the quality of the water delivered from a watershed. In general it should be carried out in such a way that the water falling upon the clean portions of the watershed is not obliged to pass through the swamp before entering the reservoir. This may be accomplished by a system of marginal drains or canals. The lowering of the water-table of a swamp also improves the quality of the water delivered from it.

Small mill-ponds and other imperfectly cleaned ponds or pools are also frequent breeding-places of microscopic organisms. Again the Boston water-supply furnishes an example. A short distance above Basin 3 there were at one time several mill-ponds. These ponds were favorite habitats of Synura. These organisms were often found there in large numbers, and when the water was let down-stream through the mills or when heavy rains caused the ponds to overflow, the Synura would become numerous in Basin 3.

Thus it is seen that in order to avoid the growth of troublesome organisms the water should be delivered from a watershed *quickly*, and should not be allowed to stand in shallow ponds or pools in contact with organic matter. As far as possible a watershed should be self-draining. It may be added that the storage reservoir also should be self-draining. It often happens, when the bottom of a reservoir is uneven, that water is left in small pools as the reservoir is drawn down. These pools are usually shallow and the water becomes warm and stagnant. They often become filled with rich cultures of organisms, and when they overflow the organisms are scattered through the reservoir. Such pools or pockets should be provided with an outlet. If this is impossible it may be advisable to fill them up. The author once observed a "pocket" in a reservoir that was excavated to a considerable depth for the sake of removing all the organic matter at the bottom. This pocket could not be drained, and during the summer it became the breeding-place of Synura and other Protozoa. It would have been better to have removed a portion of the organic matter and to have covered the remainder with clean material.

It has been stated that water should not be allowed to stand for any length of time in contact with organic matter. It is quite as bad for water to stand over a swamp as it is for it to stand in a swamp. It may be worse, for if the water has sufficient depth the decomposition of the organic matter at the bottom may take place in the absence of oxygen, and under these conditions some of the resulting products are more easily taken up by the water. This brings us to the consideration of the so-called " stagnation effects."

Stagnation.—By the term "stagnation" is meant a continued state of quiescence of the lower layers of water in a lake or reservoir caused by thermal stratification, as described in Chapter V. During these periods of quiescence the water below the thermocline, i.e. the stagnant water, undergoes certain changes,—the character and amount of these changes varying with the nature of the water and especially with the presence or absence of organic matter at the bottom of the reservoir. Stagnation may be studied best in ponds where there is a considerable deposit of organic matter at the bottom, and of such ponds Lake Cochituate is an excellent example.

Near the efflux gate-house the lake has a depth of 60 ft. At the bottom there is a layer of organic matter of unknown thickness. The upper portion of this is due to deposition of organisms and other organic material transported by the water. The period of summer stagnation extends from April to November, and during this time the deposit of organic matter at the bottom is accumulating.

The changes that take place in the water at the bottom of Lake Cochituate during the summer are shown in the following table, where the analyses of the water at the surface and bottom are compared. The most conspicuous change is that of the color (see Fig. 17). While the water at the surface is bleaching under the action of the sunlight, that at the bottom grows rapidly darker until, near the close of the stagnation period, it has a decided opalescent turbidity and a

STORAGE OF SURFACE-WATER. 135

CHEMICAL ANALYSES * OF WATER AT THE SURFACE AND BOTTOM OF LAKE COCHITUATE DURING THE PERIOD OF SUMMER STAGNATION, 1891.

PARTS PER MILLION.

Date.	Temperature.		Color.		Albuminoid Ammonia.		Free Ammonia.		Nitrites.		Nitrates.		Fixed Solids.		Hardness.		Iron.		Manganese.		Silica.		Dissolved Oxygen.	
	Surface.	Bottom.	Surface.	Bottom.†	Surface.	Bottom.	Surface.	Bottom.	Surface.	Bottom.	Surface.	Bottom.	Surface.	Bottom.	Surface.	Bottom.	Surface.	Bottom.	Surface.	Bottom.	Surface.	Bottom.	Surface.	Bottom.
Apr. 3	48.6	43.2	.36	.36	.182	...	.006	...	.001	...	.300	...	...	...	18.0	...	...	.6	...	.7	...	.4	...	...
Apr. 8	...	...	...	...	...	...	...	...	...	...	...	...	25.0	27.5	...	18.0	...	.7	...	.5	5.9	3.0	Per cent of saturation.	
June 6	69.1	44.4	.33	.88	.170	.190	.016	.224	.002	.002	.200	.210	30.5	36.5	17.4	18.4	.3	1.6	.5	.6	...	4.4	...	
July 17	73.9	41.4	.21	1.51	.174	.212	.004	.430	.002	.004	.120	.080	24.5	40.0	17.6	19.5	.6	5.0	.3	.7	...	...	...	
Aug. 19	74.4	43.7	.19	2.56	.156	.262	.012	.600	.004	.006	.020	.030	21.5	51.5	18.9	21.5	1.1	9.7	.4	2.0	...	...	79	0
Sept. 28	71.2	44.3	.13	2.93	.134	.244	.009	.736	.002	.005	.020	.020	35.5	58.0	18.2	19.8	1.1	9.6	.4	2.1	3.1	8.6	100	0
Oct. 23	57.4	43.7	.16	3.75	...	.355	...	.880	...	.003	...	.000	...	64.5	...	22.0	...	13.2	...	2.9	...	...	...	
Nov. 2	43.4	44.6	.33	...	.144	...	.044	...	.001	...	.200	...	35.0	...	19.0	...	...	...	...	...	...	...	...	
Dec. 2	39.3	39.9	.33	.37	.212	...	.032	...	.003	...	.120	...	32.0	...	18.0	...	...	...	...	...	...	...	...	

* Made by Dr. F. S. Hollis. † After standing several hours.

rich brown color. A peculiarity of the water is that its color deepens rapidly after being drawn to the surface. These color phenomena are due to the presence of iron in the water. By sedimentation of iron in combination with organic matter and of ferric hydrate produced by oxidation in the upper layers, a considerable deposit of iron has been formed at the bottom. As the oxygen dissolved in the water at the bottom disappears during the summer, the ferric iron gives up its

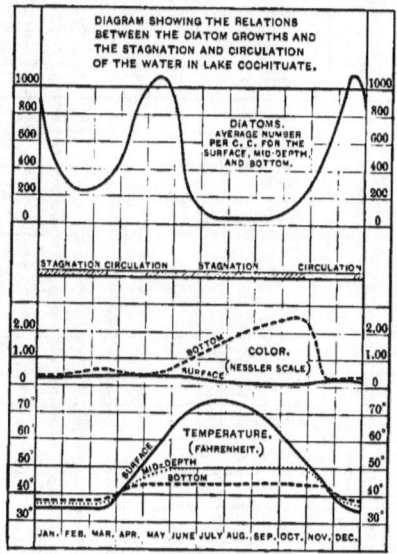

FIG. 17.—STAGNATION EFFECTS—LAKE COCHITUATE.

oxygen to the organic matter and becomes reduced to the ferrous state. In this state it is soluble. As stagnation continues it becomes dissolved in increasing amounts. When carried to the surface it becomes oxidized to the insoluble ferric state, deepening the color of the water for a time, but later precipitating as a brown sediment and leaving the water with little color. Important changes in the organic matter in the lower layers take place during the stagnation periods. The amount of organic matter in the water increases by sedimentation from above and by solution from the ooze at the bottom. The albuminoid ammonia increases. Decomposition of the organic matter takes place. The dissolved oxygen disappears and the nitrates, iron, etc., become reduced. The free ammonia and nitrites increase. After the supply of oxygen has become exhausted, putrefaction through the agency of the anaerobic bacteria takes place and the water acquires offensive odors. Increasing amounts of mineral matter are taken up from the bottom by the lower layers of water. This is true not only of iron, but also of silica, manganese, and some of the calcium and magnesium salts.

These stagnation effects are observed only below the thermocline. The relative changes that occur at different depths are well shown by the amount of dissolved oxygen, and the progress of the changes through the season may be studied by a series of such observations. The following table serves to illustrate this:*

DISSOLVED OXYGEN AT VARIOUS DEPTHS IN LAKE COCHITUATE, IN PER CENT OF SATURATION.

	Aug. 16, 1891.	Sept. 28, 1891.
Surface	79	90
10 ft.	84	81
20 "	36	33
30 "	21	9
40 "	20	8
45 "	2	—
50 "	0	0
56 "	—	0
57½ "	0	—

* Much more elaborate studies upon this subject have been made at Jamaica Pond by the Massachusetts State Board of Health. For further details the reader is referred to the Special Report of 1890 on Examination of Water-supplies, and to the Annual Reports for 1891 and 1892.

The effect of stagnation upon the microscopic organisms has been referred to. Little life exists below the thermocline. The ooze at the bottom is largely an accumulation of dead organisms. The few living organisms that are found there are Fungi, Protozoa and Crustacea,—organisms that are parasitic or that play the part of scavengers. The water at the bottom, however, acquires a supply of food-material—both organic and mineral—suitable for microscopic life. After stagnation ceases and the period of circulation begins, this food-material is carried to the upper regions where, with light and oxygen, the organisms are able to utilize it. The diatoms in particular depend upon the food-supply acquired by the water during periods of stagnation.

The stagnation of a pond that has deposits of organic matter at the bottom affects the quality of the water in two ways. When the bad water at the bottom is carried to the surface during the periods of circulation the entire body of water is affected by it. The color increases, the organic matter increases, and the odor may become unpleasant. These are the direct effects. Odors of the water that are caused by the growth of organisms that have been stimulated by the acquired food-materials are the indirect effects.

The disagreeable effects of stagnation are not dependent upon the depth of a pond, except in so far as the depth affects thermal stratification. They depend somewhat upon the character of the water stored, but much more upon the amount and character of the organic matter at the bottom and upon the length of the stagnation periods. If the bottom of the reservoir contains no organic matter the phenomena described above will not occur. It has been found that in Basin 4 of the Boston water-supply, where the organic matter was carefully removed from the bottom, the dissolved oxygen at the bottom does not become exhausted during the stagnation periods, although it is appreciably reduced in amount. The author once collected a sample from Lake Champlain at a depth of nearly 400 ft. The temperature was 39.2°—i.e. maximum density—and the water was probably in a state of permanent stagnation. The sample was bright, clear, colorless, and without odor. The material on the bottom was found to be almost perfectly clean gravel.

Organic matter at the bottom of shallow reservoirs will cause a deterioration of the water stored in them. If there is no summer stagnation the water at the bottom becomes warm, and decomposition goes on rapidly. The products of decay taken up by the water support the growth of organisms,—particularly the blue-green algæ. Moreover, during the winter when the surface is frozen these shallow ponds grow stagnant and the conditions become similar to those in deep ponds. After the periods of winter stagnation, shallow ponds often contain heavy growths of diatoms. Organic matter at the bottom of a shallow reservoir affects the quality of the water in another way. It offers support for fixed aquatic plants, and these may injure the quality of a water directly by their decay or indirectly by harboring the microscopic organisms.

The evidence is conclusive that the removal of the organic matter from the bottom of a reservoir is an important factor in the prevention of the growth of troublesome organisms. To what extent engineers are warranted in expending large sums of money for the cleaning of reservoir sites is a matter for expert opinion in every individual case. The removal of trees, stumps and vegetation is always wise. The removal of the upper layers of soil that contain large percentages of organic matter is usually advisable but not always necessary.

In lakes of the first order, where the lower layers of water are in a state of permanent stagnation, the character of the bottom has little effect upon the whole body of water because the lower layers do not become mixed with those above. In artificial reservoirs, however, this condition seldom obtains. In reservoirs designed to store water from a watershed upon which there are large swamp areas and where the water entering the reservoir is liable to contain amounts of suspended organic matter sufficient to form a considerable deposit at the bottom, it is not wise to remove the top soil to any great depth, because in the course of a comparatively few years the effect of the organic matter deposited by the water may equal that originally present in the soil. In some cases where deposits of peat, muck, etc., extend to great depths it may be found advisable to cover the surface of the organic matter with clean material rather than to excavate it. With a clean watershed, however, the best engineering practice endorses the removal of the top soil to such a depth that the upper layer of soil remaining shall contain not more than about 2% of organic matter, as determined by the loss on ignition of a sample dried at 100° C.

Wherever possible, deep storage reservoirs should be so designed that the lower layers of water may be drawn off and wasted through a low-level gate. This is especially important in reservoirs where the top soil has not been removed. It is much better to waste the bad water at the bottom caused by stagnation than to allow it to affect the entire body of water in the reservoir.

CHAPTER XI.

STORAGE OF GROUND-WATER.

Ground-water must be stored in the dark in order to prevent the growth of microscopic organisms.

Water that has passed through the soil usually carries much mineral matter in solution, some of which forms an important ingredient of plant-food. When such water is stored in an open reservoir it is liable to deteriorate. Diatoms especially are liable to develop, because their mineral contents are greater than those of most plants. These growths are less likely to occur in a new reservoir than in one that has been long in use. The seeding of the reservoir must first take place. As a rule some of the littoral organisms develop first, growing on the sides or even on the bottom of the reservoir. Gradually a deposit of organic matter collects at the bottom, and the conditions become favorable for the growth of the limnetic organisms.

Of the diatoms that occur in ground-water exposed to the light Asterionella is by far the most troublesome. Others may make the water turbid, but the Asterionella is very odoriferous. In surface-waters it has been found that this organism develops most vigorously after the stagnation periods. It is probable that this is true also in ground-waters. Most reservoirs for the storage of ground-water are shallow and of comparatively small size. Often water is not pumped directly through them. Such reservoirs become stagnant at times, and it has been observed that in them the Asterionella show a spring and fall seasonal distribution like that observed in surface-waters. It sometimes happens that for many years an open reservoir gives no trouble, but that finally a layer of organic matter accumulates at the bottom, the water in some way becomes seeded with Asterionella, and thereafter regular growths of these organisms occur. If open reservoirs are to be used for the storage of ground-water they should be kept clean.

When a water-supply is taken partly from the surface and partly from the ground it is even more necessary that covered storage reservoirs should be used, because the surface-water may contain organisms the growth of which in the reservoir may be stimulated by the food-material in the ground-water, and because organic matter will be deposited from the surface-water, increasing the effects of stagnation and making it possible for Asterionella growths to occur. The water-supply of Brooklyn, N. Y., presents an interesting example.

The supply of this city is derived from a number of small storage reservoirs along the southern shore of Long Island and from fourteen driven-well stations along the line of the aqueduct. The well-water is drawn from depths varying between 25 and 200 ft. The waters become mixed in the aqueduct and are stored in three basins comprising Ridgewood Reservoir. The different sources of water vary greatly in character. Some contain an abundance of organic matter; some have high free ammonia, nitrites, and nitrates; some have considerable iron; and one or two have high chlorine and hardness due to admixture of a small amount of sea-water. The watershed is sandy, and all the waters are rich in silica.

In 1896 Asterionella developed in Ridgewood Reservoir in great abundance, and since then it has reappeared at intervals. In a general way these growths have shown the spring and fall distribution, but they also correspond to some extent with increased proportions of ground-water used. At times the numbers of Asterionella present have been very high,— 25 000 or 30 000 per c.c. For many years Ridgewood Reservoir caused no trouble and the water-supply bore an enviable reputation. It was not until a considerable deposit of diatoms and other organic matter had accumulated on the bottom of the basins and until the amount of ground-water had come to be about 40% of the total supply that the conditions became favorable for such enormous growths of Asterionella. Fortunately for the consumers, a by-pass around the distributing-reservoir permits the water to be pumped from the aqueduct directly into the distribution system. This is used whenever the Asterionella in the reservoir become abundant enough to cause a bad odor.

Water that has been filtered is practically a ground-water. Modern filter-plants therefore often provide that the filtered water shall be kept in the dark until used. The filter-beds themselves are often covered. This is chiefly to prevent freezing during the winter, but it also serves to prevent growths of algæ upon the surface of the sand. These growths are sometimes very thick and add considerably to the cost of filtration by requiring more frequent scraping of the filter-beds.* Of the organisms that grow upon the surface of sand-filters

---

* A growth of organisms upon the surface of a sand-filter sometimes has its advantages. On the filter-beds at Far Rockaway, L. I., where a ground-water supply is aerated and filtered for the sake of removing the iron, an extensive growth of confervoid algæ and filamentous diatoms develops during warm weather. When the water is drawn off preparatory to scraping the filter, these algæ growths form a fibrous layer upon the surface of the sand. This matting is so tough that it may be rolled up in sheets, and as it contains a large percentage of the iron removed from the water it materially reduces the labor of scraping the sand, and has the further advantage that it removes comparatively little sand.

Tabellaria and Spirogyra are perhaps the most important, but a great variety of both animal and plant forms is often observed in the water above the sand. The extent of the growth of organisms at the surface of the sand may be illustrated from the records of an experimental filter. After running continuously for 25 days the sand became so clogged that scraping was necessary. A microscopical examination of the surface showed that over each square centimeter there were 2 500 000 Tabellaria and 1 000 000 Synedra, besides many other organisms in smaller numbers. Calculation from the analyses of the water filtered showed that during the 25 days 150 000 Tabellaria and 20 000 Synedra were removed from the water by each square centimeter of the filter. The difference between the two sets of figures represents the growth of the organisms upon the sand.

Darkness is not always sufficient to prevent a ground-water from deteriorating. There are some organisms that can live without light, and indeed prefer darkness. Of such a nature are the fungi (using the word in its broad sense as including those vegetable forms destitute of chlorophyll) and some of the Protozoa.

Crenothrix is the most important organism of this character that affects ground-water supplies. It is a small filamentous plant the cells of which are but little larger than the bacteria. Its filaments have a gelatinous sheath colored brown by a deposit of ferric oxide. It grows in tufts, sometimes matted together into a felt-like layer.

Crenothrix is liable to occur in ground-water rich in iron and organic matter. It frequently infests water obtained from wells driven in swampy land. It is often observed in imperfectly filtered water. It may grow in almost any part of the system,—in the driven wells, filter-galleries, reservoirs, and distribution-pipes. It is especially liable to occur about woodwork.

Crenothrix causes trouble in tubular wells by choking them with deposits of iron. Leptothrix, Spirochætæ, and allied organisms also do this. Crenothrix causes trouble in the service-pipes by reducing the capacity of the pipe. But it causes most trouble when the filaments break off and become scattered through the water. It is then liable to make the water unfit for laundry use on account of deposits of iron-rust.

Crenothrix has caused annoyance in many water-supplies. The "water calamity" in Berlin first drew attention to its evil effects. In 1878 the water from the Tegel supply became filled with small, yellowish-brown, flocculent masses which settled to the bottom when the water was allowed to stand in a jar. The odor of the water and the effects of the iron oxide in washing were decidedly troublesome. Crenothrix was not found in Lake Tegel, but was found in many wells, in the reservoirs at Charlottenburg and in the unfiltered water of the river Spree.

In 1887 the water-supply of Rotterdam was badly affected with Crenothrix. The water was drawn from the river Maas, and, after sedimentation, was filtered. At the time when Crenothrix appeared the system was being enlarged. New filter-beds were in use, but the filtered water was conducted through the old conduits and the old reservoir to the old pumps. In the old conduit, or flume, there were many wooden timbers, and on these Crenothrix was found growing in abundance. Inspection showed that some of the water was imperfectly filtered, and that this impure water was the chief cause of the sudden and extensive development of Crenothrix.

CHAPTER XII.

GROWTH OF ORGANISMS IN WATER-PIPES.

THE reactions between the water and the water-pipes of a water-works system involve such matters as iron-rusting, tuberculations, lead-poisoning, and others of a chemical and physical nature. There are also biological reactions. These may be considered under two heads: (1) the effect of the aqueducts and pipes upon the biology of the water, and (2) the effect of the water upon the biology of the aqueducts and pipes.

1. The temperature of water changes during its passage

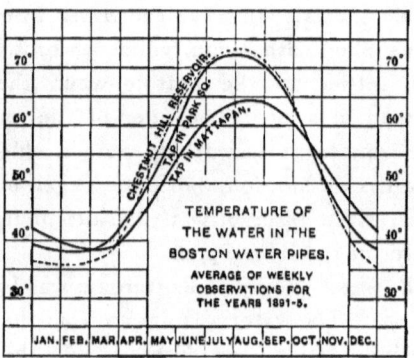

FIG. 18.

through the pipes of a distribution system. The nature of these changes is shown by Fig. 18, where the curves represent the averages of weekly temperature observations for five years at Chestnut Hill Reservoir and at two taps, one at Park Square, 5 miles from the reservoir, and the other at Mattapan, 11 miles from the reservoir. During the spring and summer the water grows cooler as it passes through the pipes, and during the autumn and winter it grows warmer. The maximum temperature at Mattapan is never as high as that at Park Square, but the minimum temperature is about the same at both places, though it occurs later in the season at Mattapan.

Samples taken at the same places serve to illustrate the changes that take place in the organisms of the water due to their passage through the pipes. Weekly observations for five years (1891–5) showed the following average number of organisms present:

	Number of Standard Units per c.c.	
	Organisms.	Amorphous Matter.
Chestnut Hill Reservoir	248	222
Brookline Reservoir	215	212
Tap in Park Square	189	190
Tap in Mattapan	81	105

The greatest reduction does not occur near the reservoirs, where the pipes are large and the currents swift, but at the extremities of the distribution system, where the pipes are smaller.

The observations showed that during the winter, when there are comparatively few organisms in the water, the reduction in the pipes is much less than during the summer, when organisms are more abundant. During the six months of the year, from November to April, there was a reduction of 44% in organisms and 24% in amorphous matter in about 6 miles of pipe; while during the six months from May to October the reduction was 62% for the organisms and 53% for the amorphous matter. It is worth noting that the reduction in organisms was greater than the reduction in amorphous matter.

Not only are the microscopic organisms and amorphous matter reduced in the pipes, but the bacteria also tend to decrease. This fact has been observed in many cities. In the pipes of the Boston Water Works the decrease does not occur throughout the entire year. In the summer, when the temperature of the water is high and when the organisms in the water and those growing in the pipes are passing rapidly through stages of growth and decay, there is a considerable increase. This is shown in Fig. 19.

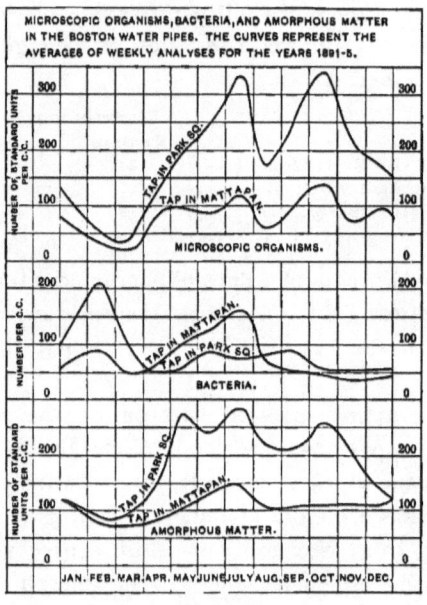

FIG. 19.

In order to determine what organisms showed the greatest reduction in the pipes, a detailed study of the examinations above referred to was made for the years 1892 and 1893. The following were the results:

PERCENTAGE REDUCTION OF MICROSCOPIC ORGANISMS IN THE DISTRIBUTION-PIPES BETWEEN PARK SQUARE AND MATTAPAN, BOSTON, MASS.

	Average for the years 1892 and 1893.
Diatomaceæ	58 per cent
Chlorophyceæ	57 " "
Cyanophyceæ	54 " "
Protozoa	64 " "
Miscellaneous	58 " "
Organisms of all kinds	56 " "

Questions naturally arise as to the cause and effect of this reduction of organisms in the pipes. They may be considered under the following topics: sedimentation, disintegration, decomposition, and consumption by other organisms.

Most of the microscopic organisms are heavier than water. Some always settle in quiet water, and they do so in the pipes whenever the current is reduced to a certain point. Others, which in ponds usually rise to the surface on account of the gas-bubbles which they contain, will settle in the pipes when the pressure of the water has deprived them of their gas. In dead ends the organisms and particles of amorphous matter often accumulate and form deposits upon the bottom of the pipes. They also tend to deposit on up-grades. It is a matter of frequent observation that the water from the high points of a distribution system contains fewer organisms than that from the low points. The same fact has been observed in high buildings, where the difference between the water on the upper stories and that on the lower floor is often considerable.

Many of the common organisms are very fragile. Even a slight agitation of the water will break them up. This is particularly true of certain Protozoa, but it also happens to the siliceous cells of diatoms.

The organisms found in surface-waters are accustomed to live in the light. When they enter the dark pipes they are liable to die and decompose. This is particularly true of some of the organisms that are abundant in the summer. Microscopical examination of samples from the service-taps has often revealed organisms in a decomposing condition, swarming with bacteria. This decomposition tends to reduce the numbers of organisms in the pipes.

Another important consideration in the reduction of organisms is the fact that in many of the distribution systems where surface-waters are used the pipes are covered with growths of sponge, etc. These attached growths depend for their food-material upon the minute organisms found in the water. If the growths are abundant, the removal of organisms from the water by this means may be considerable.

2. Comparatively little has been written in this country upon the biology of aqueducts and pipes. Our attention has been called to growths of Crenothrix and of fresh-water sponge, but no attempt has been made to give an accurate account of the organisms infesting the distribution systems of our water-supplies. In Europe, however, the subject has been considered to some extent.

In the city of Hamburg the minute animals inhabiting water-pipes were studied by Hartwig Petersen in 1876. Ten years later Karl Kraepelin made a more extended study. His observations were of much interest. He found an animal growth, often more than one centimeter thick, covering the entire surface of the pipes. The composition of this growth varied in different places. He gave a list of sixty different species observed. In many places the walls of the pipes were covered with fresh-water sponges, chiefly Spongilla fluviatilis and Spongilla lacustris. Mollusks were conspicuous, especially the mussel, Dreyssena polymorpha. Snails were also numerous. Hundreds of "water-lice" (Asellus aquaticus) and "water-crabs" (Gammarus pulex) were found at every examination. The material known as "pipe-moss" was common, and consisted largely of Cordylophora lacustris and the Bryozoa, Plumatella and Paludicella.

At the time when Crenothrix was giving so much trouble at Rotterdam, Hugo de Vries made an extended study of the animals and plants found in the water-pipes of that city. His observations were confined chiefly to the pipes and canals which conveyed the unfiltered water of the river Maas to the filter-beds. In speaking of one of the canals he said: "The walls were thickly covered with living organisms up to the water-level. They formed an almost continuous coating of varying composition. There were only one or two exceptions to this. In one place, where the water came from the pumps with great velocity, the walls were free from living organisms; and in another place, where there was almost no current, only one living form was seen. There was a section of one of the canals, where a gentle current was flowing, that was a magnificent aquarium. The walls were everywhere covered with white tufts of fresh-water sponge (Spongilla fluviatilis). Many of these tufts reached a diameter of 6 or 8 inches, but most of them were somewhat smaller. Between the sponge patches were seated countless numbers of the mussel, Dreyssena polymorpha. Individuals old and young were often seen grouped together in colonies which sometimes extended completely over the sponges. But what most of all attracted attention was a luxuriant growth of the 'horn-polyp,' Cordylophora lacustris. It covered the mussel-shells and occupied all the space between the sponges. The stalks reached a length of an inch or more. On and between the Cordylophora swarmed countless numbers of Vorticella, Acineta, and other Protozoa and Rotifera. These organisms had no lack of food-material, and the absence of light protected them from many foes which, in the light, thin out their ranks. Over all these animals Crenothrix was found growing in abundance. The shells of the mussels and the stems of the 'horn-polyps' were coated with a thick felt-like layer of these 'iron-bacteria.' In other localities in the pipes the place of the 'horn-polyps' was occupied by the Bryozoa, or 'Moss-animalcules.' All of these branching forms were spoken of collectively by the workmen as 'pipe-moss.'"

In the summer of 1896, when the pipes of the Metropolitan Water Works were being laid in Beacon Street, Boston, near the Chestnut Hill reservoir, a 16-inch main leading from the Fisher Hill reservoir to the Brighton district was opened. This afforded an opportunity to examine the material on the inside of a pipe that had been laid ten years. Inspection showed that besides the usual coating of iron-rust, tubercles, etc., there were numerous patches of fresh-water sponge (both Spongilla and Meyenia), brownish or almost white in color, and about the size of the palm of one's hand. What was most conspicuous, however, was a sort of brown matting which covered much larger areas, and which had a thickness of about $\frac{1}{4}$ inch. It had a very rough surface and, when dried, reminded one of a piece of coarse burlap. This proved to be an animal form belonging to the Bryozoa, known as Fredericella. As fragments of it had several times before been observed in the water from the service-taps, and as it had been seen growing in some small pipes connected with the filtration experiments at the Chestnut Hill reservoir, more extended observations were made in different parts of the distribution system.

These brought out the fact that sponges and the Bryozoa were well established in the pipes. Many other organisms were also observed. In some places almost pure cultures of Stentor and Zoothamnium were found. At other points hosts of different organisms were seen, such as snails, mussels, Hydra, Nais, and Anguillula, Acineta, Vorticella, Arcella, Amœba, countless numbers of ciliated infusoria, and many other forms. The growths were distinctly animal in their nature, but in many places parasitic vegetable forms, such as Achlya, Crenothrix, Leptothrix, etc., were common. The most important class of organisms found, however, was the Bryozoa, of which Fredericella and Plumatella were the chief representatives.

The fact that the organisms that dwell in water-pipes depend for their food-material upon the algæ, protozoa, etc., contained in the water may be easily demonstrated by experiment. Specimens of Fredericella and Plumatella were once placed in a series of jars, some of which were supplied with water rich in its microscopic contents, while others were supplied with the same water after filtration. All the jars were kept in semi-darkness at the same temperature, and were examined daily. The Fredericella and Plumatella that had been supplied with filtered water soon began to die, while those in the other jars lived as long as the experiment was continued. Some of the same Bryozoa were placed in jars furnished with water from the Newton supply,* and after

---

* A ground-water almost free from microscopic organisms.

about a week they died for want of food. Dr. G. H. Parker* once made a similar experiment on fresh-water sponge, and obtained the same result. With these facts established, we may confidently affirm that fresh-water sponge, Bryozoa, and similar pipe-dwellers will be absent from water-pipes where ground-water or water that has been effectively filtered is used.

One naturally asks, "What is the effect of these organisms growing in the pipes?" In a certain sense they tend to improve the quality of the water by reducing the number of floating microscopic organisms; but they themselves must in time decay, and any one whose nose has ever had an experience with decomposing sponge will appreciate the fact that better places for these organisms may be found than the distribution systems of our water-supplies. It should be stated, however, that in all probability very large quantities would be required to produce tastes or odors that would be noticed in the water. Perhaps the greatest objection to their presence is the fact that they tend to impede the flow of water in the pipes. When one considers that a coating $\frac{1}{4}$ inch thick diminishes the area of the cross-section of a 24-inch pipe by 4%, and of a 6-inch pipe by 15%, and when one learns that these organisms often form layers even thicker than this, it will be seen that such growths are matters of no little importance. Furthermore, fingers of the fresh-water sponge sometimes extend several inches into the water, and the matting of the Bryozoa is always rough on account of the stiff branches that are extended in order that the organisms may secure their food. This roughness of the surface must

---

* G. H. Parker, Experiments on Fresh-water Sponge, Special Report of the Massachusetts State Board of Health, 1890, p. 618.

increase the friction of the pipe by a considerable but indefinite amount.

An interesting experience with pipe-moss is on record at the Brooklyn Water Department. In November, 1897, the water in the Mt. Prospect reservoir became so filled with Asterionella that it was deemed advisable to shut off the reservoir and pump directly into the pipes. This action was followed by the appearance of brown fibrous masses in the tap-water. In a number of instances this fibrous matting stopped up the taps, and even large pipes were choked. The water at the same time had a distinctly moldy and unpleasant odor. The fibrous matting proved to be Paludicella. It had been growing on the inner walls of the pipes, and the change of currents and the pulsations of the pump, due to the direct pumping into the pipes, had dislodged it. Systematic and thorough flushing of the pipes materially improved the conditions.

PART II.

CHAPTER XIII.

CLASSIFICATION OF THE MICROSCOPIC ORGANISMS.

The microscopic organisms found in drinking-water include the lowest forms of life. Some of them belong to the vegetable kingdom, some belong to the animal kingdom, while others possess characteristics that pertain to both. There is in reality no sharp dividing-line between the vegetal and the animal in the low forms of life. Nature's boundaries are always shaded on both sides.

Classification of organisms into groups is necessary, but it must be borne in mind that all classifications are artificial and subject to change. The one outlined below and used throughout this volume is believed to be the most convenient for the work at hand. A number of groups, not pertaining to the microscopical examination of drinking-water, are omitted.

CLASSIFICATION OF THE MICROSCOPIC ORGANISMS.

Plants.	Animals.
DIATOMACEÆ.	PROTOZOA.
SCHIZOPHYCEÆ.	*Rhizopoda.*
Schizomycetes.	*Mastigophora (Flagellata).*
Cyanophyceæ.	*Infusoria* (in the narrower sense).
ALGÆ (in the narrower sense).	ROTIFERA.
Chlorophyceæ.	CRUSTACEA.
FUNGI.	*Entomostraca.*
VARIOUS HIGHER PLANTS.	BRYOZOA (POLYZOA).
	SPONGIDÆ.
	VARIOUS HIGHER ANIMALS.

CHAPTER XIV.

DIATOMACEÆ.

The Diatomaceæ comprise a group of minute vegetable forms of a low order. Their exact position in the scale of life has been the subject of much controversy. The early writers considered them to belong to the animal kingdom because of the power of movement that some of them possess. Later, when they had become generally recognized as plants, they were considered as a Class or Order of the Algæ. Recent cryptogamists, however, prefer to class them as an independent group, thereby recognizing the fact that they are quite different from most unicellular plants. This difference lies chiefly in the possession of siliceous cell-walls upon which may be observed certain markings that are constant in size and arrangement for each species. The great beauty of these markings, together with the infinite variety in the sizes and shapes of the cells of different species, have long made them objects of special study by microscopists.

Diatom Cells.—A diatom cell is constructed like a box. There is a top and a bottom, known as the upper and lower valve, on both of which markings are found. The valves are connected by membranes known as "sutural zones," "connective membranes," "girdles," or, when detached, as "hoops." There are two of these membranes, one attached to each valve, and they are so arranged that one slides over the other just as the rim of a box-cover fits over the sides. This arrangement may be seen in Plate I, Figs. A, B, and C, where a typical diatom, Navicula viridis, is shown in three views. A represents the valve* view of the diatom, that is, the view seen when looking directly at the valve or the top of the box. B represents the girdle* view, the view seen when looking at the connective membrane. C is a cross-section through the diatom.

The upper or outer valve is indicated by a, and its connective membrane by c. The girdle view shows how this connective membrane of the larger valve fits over a similar one, c', attached to the lower or smaller valve, b. These girdles have the power of sliding one upon the other so that the thickness of the diatom, i.e. the distance between the valves, is variable.

The valves of the diatom shown in the figure are covered with furrows or markings, g. At the centre and at each end there are slight thickenings of the cell-wall, known as nodules. The central one is called the central nodule, d, and those at

---

* The terms used by different writers to express these two views of a diatom are very confusing. In the following list the terms under A represent the valve view and those under B the girdle view.

A	B
Valve view.	Girdle view.
Side view.	Front view.
Top view.	Zonal view.
Face valvaire.	Face connective.
Primary side.	Secondary side.
Secondary side.	Primary side.
Vue de profil.	Vue de face.

The terms "side view" and "front view" are those generally used by English and American diatomists, but the author has avoided them as not being in themselves sufficiently clear, and has preferred to use the less euphonious but more self-explanatory terms, "valve view" and "girdle view." In consulting books on diatoms the reader should be careful to note the way in which the two views are designated.

the ends, terminal nodules, *e, e*. Between these nodules and extending along the medial line of the valve there is a sort of ridge, *f*, in which there is a furrow called a raphe, or raphé. Through this the living matter of the diatom probably communicates with the outer world. The slit is supposed to be somewhat enlarged at the nodules. The raphé, the nodules, and the markings, taken in connection with the shape and size of the valves, are the most important external features of a diatom and are the first to be considered in studying them.

Shape and Size.—There is probably no class of unicellular organisms in which the outlines vary more than in those of the diatoms. From the straight line to the circle almost all the geometrical figures may be found. Some of these may be described as circular, oval, oblong, elliptical, saddle-shaped, boat-shaped, triangular, undulate, sigmoid, linear, etc. The variations in shape are most marked in the valve view. The girdle view, as a rule, is more or less rectangular. The valves are usually plane surfaces, with only slight curvatures or undulations. Occasionally the surface is warped as in Amphiprora and Surirella. As a rule the two valves of a frustule are nearly parallel, but in such forms as Meridion, Gomphonema, etc., the frustule is wedge-shaped when seen in girdle view. The most varied forms are found in salt or brackish water, and the common fresh-water forms are so simple and so characteristic that the reader will have little difficulty in assigning them their proper generic names. Some genera have the cell divided more or less completely by internal plates, called septa, when fully developed as in Rhabdonema; and vittæ, when incomplete as in Grammatophora. Some diatoms have external expansions on the margin of the valves. Surirella, for example, has thin expansions known as alæ, or wings. When these alæ are imperfectly developed they are called keels. Nitzschia for this reason is said to be carinate. These wings or keels usually extend along the border of the raphé. Certain filamentous forms, such as Melosira, have processes at the point of attachment. In others these processes are elongated into horns, or bristles.

Diatoms vary in size from the minute Cyclotella, less than 10 microns * in diameter, to such large forms as Surirella and Navicula, that sometimes are one millimeter long. Some filamentous forms grow to a considerable length,—often several feet.

Markings.—The valves of most diatoms are marked with lines or points. In many cases the lines may be resolved into series of points, pearls, beads, or striæ, when a higher power of the microscope is used. The variations in the number and size of these points and their uniformity in different individuals of the same species make them convenient objects for testing the resolving power of microscopes. The variation in the number of these striæ may be seen from the following table:

	Number of Striæ per Millimeter.	
	Longitudinal.	Transverse.
Epithemia ocellata, Kz.	800	430
Navicula major, Kz.	850	630
" viridis, Kz.	2400	720
" lyra	850	1000
Cymbella navicula, Ehb.	1200	1500
Pleurosigma angulatum, Sm.	1580	2100
Synedra pulchella, Kz.	670	2150
Navicula rhomboides	1700	2700
Amphipleura pellucida, Ktz.	3400	3700 to 5200

The extreme minuteness of these points, their various appearances under different conditions, and the difficulty of studying them even with microscopes of the highest magnifying powers, have given rise to many different theories concern-

---

* One micro-millimeter, or micron (u), equals .001 millimeter.

ing the character of the valves. Some writers insist that the points are elevations: others claim that they are depressions. Recent students agree that the structure is more complex than was formerly considered to be the case. The following conception of M. J. Deby, while perhaps not correct for all cases, is a good illustration of the modern view (see Fig. 20).

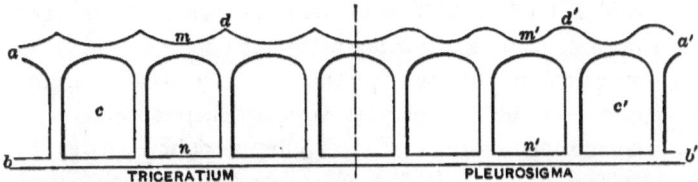

FIG. 20.—TRANSVERSE SECTION OF A DIATOM VALVE. (After Deby.)
a. Upper (outer) layer.
b. Lower (inner) layer.
c. Cavities.
d. Inter-alveolar pillars.
m. Thin part of upper layer.
n. Bottom of alveolæ.

"The valves of most diatoms are composed of two layers, between which there are circular or hexagonal cavities bounded by walls of silica. The upper layer is not uniform in thickness, but is thin just above the cavities, and thicker, rising in pointed or rounded prominences, above the intersection of the walls of the cavities. The upper layer is lightly silicified, and the thin portions are easily broken, making openings into the cavities. The lower layer bears varied designs the nature of which has not been well established. What authors have described as areolæ, pearls, pores, orifices, granular projections, depressions, hexagons, beads, points, etc., are really one and the same thing."

Cell-contents.—The frustule of a diatom is somewhat analogous to the shell of a bivalve,—the living matter is inside. Just inside the cell-wall there is a thin protoplasmic lining (primordial utricle). This protoplasm sends radiating streams through the cell, and it is possible that a portion of it extends through the openings in the cell-wall and communicates with the outer world. It is this layer of protoplasm also that secretes the silica of the cell-wall. Between the streams of protoplasm (Pl. I, Fig. C) there are what appear to be empty cavities. In or on the borders of these, oil-globules may be sometimes observed. There is a nucleus, and probably a nucleolus, located near the centre of the cell. The most conspicuous portion of the cell-contents, however, consists of colored lumps or plates, which are usually constant in appearance and position for any particular species. The brown coloring matter of these "chromatophore plates" is known as diatomin. It is a substance analogous to chlorophyll and has been considered by some writers to be a compound of chlorophyll and phycoxanthin. The spectrum of diatomin is very similar to that of chlorophyll. There are two absorption-bands,—one between B and C in the orange-yellow, and one between E and F in the indigo-violet. Diatomin is soluble in dilute alcohol, giving a brownish-yellow solution that is sometimes very slightly fluorescent. When dried or treated with concentrated sulphuric acid it assumes a green color. When living diatoms are exposed to the direct rays of the sun or subjected to heat for a considerable time the color of the chromatophore plates changes from brown to green. In certain species other internal features have been noted; namely, the contractile zonal membrane, the germinative dot, double nucleus, etc., but of these there is little known.

External Secretions.—Living diatoms are covered with a transparent gelatinous envelope, which is probably a secretion from the protoplasm. In many species it is very thin and can be discerned only by the use of staining agents. In the filamentous and chain-forming species it serves to hold the frustules together. In Tabellaria, for example, little lumps of the gelatinous substance may be seen at the corners of the frustules at the point of attachment. Some species secrete great quantities of gelatinous material and are entirely embedded in it. In a few cases it is of a firmer consistency and forms tubes, stalks, or stipes, upon the ends of which the frustules are seated. These stalks attach themselves to stones, wood, etc., immersed in the water.

Movement.—Some of the diatoms exhibit the phenomenon of spontaneous movement. This has always excited interest and has been the subject of much speculation. It was the chief argument advanced by the early writers for placing the diatoms in the animal kingdom. The most peculiar movement is that of Bacillaria paradoxa, whose frustules slide over each other in a longitudinal direction until they are all but detached, and then stop, reverse their motion, and slide backwards in the opposite direction until they are again all but detached. This alternate motion is repeated at quite regular intervals. Some of the free species show the greatest movement, and of these Navicula is one of the most interesting. Its motion has been described as "a sudden advance in a straight line, a little hesitation, then other rectilinear movements, and, after a short pause, a return upon nearly the same path by similar movements." The movement appears to be a mechanical one. The diatoms do not turn aside to avoid obstacles, although their direction is sometimes changed by them. The rapidity of their motion has been calculated to be "400 times their own length in three minutes." Their motion shows the expenditure of considerable force. Objects 50 or 100 times their size are sometimes pushed aside.

Various hypotheses have been advanced to account for the

movement of diatoms. Naegeli suggested that it was due to endosmotic and exosmotic currents; Ehrenberg claimed that the movement was due to cilia; another writer, that it was caused by a snail-like foot outside the frustule; another, that it was due to a layer of protoplasm covering the raphé. H. L. Smith, after much study, came to the conclusion " that the motion of Naviculæ is due to injection and expulsion of water, and that these currents are caused by different tensions of the internal membranous sac in the two halves of the frustule."

In spite of all the study that has been given to the subject, we must admit that the cause of the movement of diatoms is unknown. The "cilia theory" seems the most probable, but it is doubtful if the cilia are more than mucous threads.

Multiplication.—Diatoms multiply by a process of halving or splitting, the Greek word for which gives rise to the name *diatom*. The cell-division is similar to that in all plants, but in this case the process is of especial interest because of the rigid character of the cell-walls.

The process begins by a division of the nucleus and nucleolus. The protoplasm expands or increases in bulk, forcing the valves apart, the hoops sliding one out of the other. The two halves of the nucleus separate, the diatomin collects at either side, and a membrane forms, dividing the cell into two parts. Finally the two parts separate. The newly formed membrane becomes charged with silica, making a new valve, and soon after its hoop develops. This process is well illustrated by a drawing of M. J. Deby, shown on Pl. I, Figs. D, E, and F. Sometimes the frustules separate entirely; sometimes they remain attached forming filaments, as in Melosira, bands as in Fragilaria, or zigzag chains as in Tabellaria.

The above is the usually accepted theory of cell-division. It is probably correct in many, if not in most cases. It assumes that the siliceous walls are not able to expand, and the result is that after repeated division the frustules become smaller. It is claimed that in some cases the cell-wall does expand, and therefore that the size of the frustules does not decrease after division.

The generally accepted theory of cell-division assumes that a diatom frustule has two valves, one the larger and older, and the other the smaller and younger. After division two cells are formed, one equal in size to the larger valve and the other equal to the smaller one, the difference in size being twice the thickness of the hoop. This theory also assumes that both the mother- and the daughter-cell have the power of further division. From these assumptions certain laws of multiplication may be deduced. For example: If A is the parent cell,

After one period of time, t, A will have produced B;
" two periods " " $2t$, A " " " B',
 and B " " " C;
" three " " " $3t$, A " " " B'',
 B " " " C'',
 B' " " " C',
 C " " " D;
and so on.

From this it happens that

After t we have $1A + 1B$;
" $2t$ " " $1A + 2B + C$;
" $3t$ " " $1A + 3B + 3C + D$;
and so on.

The laws may be expressed mathematically as follows:

1. As the number of periods of division increases in arithmetical progression the total number of frustules increases in geometrical progression.

2. The number of frustules equal in size after any period of division are represented by the terms of the binomial theorem $(a + b)^n$, where a and b are unity.

These laws have been demonstrated experimentally, the first by the author* and the second by Miquel.†

Reproduction.—The continued process of multiplication results in a constant diminution in the size of the frustules. After a certain minimum limit of size has been reached or after their power of multiplication has become exhausted, a reproductive process takes place. Usually this consists of a conjugation which results in the formation of a large cell, or auxospore, capable of reproducing a frustule of large size which, by multiplication, gives rise to a new series of frustules like the first. This theory, known as "Pfitzer's Auxospore Theory," was advanced in 1871. Count Castracane has shown that its application is not universal, and that in the case of some diatoms reproduction takes place through the formation of spores, or "gonids," which become fertilized by conjugation and, after a period of repose, attain a condition for living an independent life and reproducing in every respect the adult type of mother-cell.

There are few reliable data to be found in regard to the reproduction of diatoms. True conjugation has been observed in comparatively few genera. It is believed that there are four methods of conjugation. *First*, a single frustule, self-fertilized, producing one sporange and one auxospore; *second*,

---

* G. C. Whipple, "Some Observations on the Growth of Diatoms in Surface Waters." Tech. Quar., vol. VII., No. 3. Oct. 1894.
† P. Miquel, Annales de Micrographie, No. 11, 1892.

a single frustule, self-fertilized, producing two sporanges and two auxospores; *third*, two conjugating frustules, with undifferentiated endochrome, producing one sporange and one auxospore; *fourth*, two conjugating frustules, with differentiated endochrome, producing two sporangial cells, one of which is sometimes abortive. Good examples of conjugation may be found in Surirella splendida, Epithemia turgida, and in various species of Melosira. The sporangial frustules of Melosira (shown in Pl. III, Fig. 17) are quite common.

Classification of Diatoms.—Several methods of classification of diatoms have been proposed, but only two are worthy of attention, and even these must be considered as provisional.

The most recent is that proposed by Pfitzer and elaborated by Petit. It is based upon two assumptions,—namely, that the internal disposition of the endochrome is constant for all individuals of the same species, and that the relation between the frustule and the endochrome is fixed and common to all species of the same genus. The family Diatomaceæ is divided into two sub-families, the Placochromaticeæ and the Coccochromaticeæ. The genera of the first sub-family have the endochrome arranged in plates or layers, and those of the second sub-family, in lumps or small granular masses. Secondary classification into tribes, etc., depends upon the symmetry of the valves with reference to the axes, the dissimilarity of the valves of a single frustule, the presence or absence of an intervalvular diaphragm, the raphé, nodules, etc. There is little to be said in favor of this system, but it is worthy of study as the authors have tried to do what has been long neglected,—namely, to emphasize the study of the entire cell with its contents rather than to confine the attention wholly to the cell-wall or frustule.

The most useful system of classification and the one generally recognized is that suggested by H. L. Smith. It is based almost entirely on the morphology of the frustule. This has the advantage of enabling one to classify both living and fossil forms, but it has tended to divert observers from the study of the diatom as a living cell to the study of the shell alone.

According to Smith's classification the Diatomaceæ are divided into three tribes characterized by the presence or absence of a raphé. An outline of this classification, together with descriptions of the genera most common in drinking-water, is given below. The names of the genera are printed in heavy type.

TRIBE I. RAPHIDIEÆ.

Always possessing a distinct raphé on one or both valves. Central nodule generally present and conspicuous. Frustules mostly bacillar in valve view; sometimes broadly oval; without spines or other processes. Navicula major is the typical form.

FAMILY CYMBELLEÆ.—Raphé mostly curved. Valves alike, more or less arcuate, cymbiform.

Amphora.
Frustules single, ovoidal in girdle view, the girdle often striated or longitudinally punctate. Valves extremely unsymmetrical, with a convex and concave side, with an eccentric raphé, with medial and terminal nodules. The raphé is sometimes near the convex side, sometimes near the concave side, and the medial nodule is often away from the centre. There are transverse striæ, radiating somewhat from the medial nodule. This genus is very ornate. There are a number of species, none of them very common in water. (Pl. I, Figs. 1 and 2.)

Cymbella.
Frustules generally single, elongated, symmetrical with respect

to the minor axis. Valves more or less arched, with one side very convex and the other side slightly or not at all convex; asymmetrically divided by a curved raphé; possessing terminal and medial nodules; marked by transverse bead-like striæ, which do not extend to the raphé, but have a clear space, wider at the medial nodule than elsewhere. There are a number of common species. (Pl. I, Figs. 3 and 4.)

Encyonema.
Frustules, when young, enclosed in a hyaline mucilaginous tube, in which they multiply by division, pushing each other forward in an alternately inverse position. Valves symmetrical with respect to the minor axis, convex on one side, straight on the other, with rounded extremities that project beyond the straight side. A straight raphé divides the valves into two unequal parts. There are medial and terminal nodules. The striæ are transverse or radiating somewhat from the medial nodule. There is a clear space around the medial nodule, but elsewhere the striæ approach closely to the raphé. There are several species. (Pl. I, Fig. 5.)

Cocconema.
Frustules, when young, borne singly or in pairs on filamentous pedicels, which may be simple or branched. They form mucilaginous layers on submerged objects. Later they become free-swimming. The valves are long, large, strongly arched, convex on one side, concave on the other side save for a little inflation in the middle. The raphé is curved. There are medial and terminal nodules. The striæ are rather large pearls, transverse, with very slight radiation, and not approaching the raphé closely. (Pl. I, Fig. 6.)

FAMILY NAVICULEÆ.—Valves symmetrically divided by the raphé. Frustules not cuneate or cymbiform.

Navicula.
Frustules single, symmetrical with respect to both axes. Valves naviculoid, or boat-shaped; of various proportions, some very long and narrow, others short and wide, others ellipsoidal; with straight or slightly curving sides; with ends pointed or rounded. There is a straight raphé with conspicuous medial

and terminal nodules. The valves are marked with transverse furrows, that have a slight radial tendency. The frustules are rectangular in girdle view and show the nodules plainly. There is a vast number of species and varieties, many of which are very common. In some species the striæ can be resolved into pearls. These are the Naviculæ proper. In other species they cannot be resolved, and the valves usually have wide rounded ends. These were formerly set apart as a separate genus,—Pinnularia. (Pl. I, Figs. 7 and 8.)

Stauroneis.

Frustules similar to those of Navicula. Valves symmetrical, possessing a straight raphé, with medial and terminal nodules. The striæ are pearled. There is a narrow clear space along the raphé and a wider transverse clear space at the medial nodule extending to the sides of the valve, so that the valves have the appearance of being marked with a cross. A number of species have been described, but in some instances they are very similar to Navicula. (Pl. I, Figs. 9 and 10.)

Schizonema.

Frustules quite similar to those of Navicula, and enclosed in mucilaginous tubes, as Encyonema. Raphé straight, sometimes showing a double line. Striæ generally parallel, reaching to the raphé, but not to the central nodule, around which there is a clear space. More common in salt water than in fresh water.

Pleurosigma.

Frustules like those of Navicula, but with axis turned like a letter S. Raphé sigmoidal. Striæ ornate, pearled, very fine on some species. Endochrome in two layers. (Pl. I, Fig. 11.)

FAMILY GOMPHONEMEÆ.—Valves cuneate ; central nodule unequally distant from the ends.

Gomphonema.

Frustules borne on pedicels more or less branched. Valves wedge-shaped, with more or less undulating margins and rounded ends. A central nodule near the large end. Raphé straight, dividing the valve symmetrically. Striæ pearled, transverse, radiating slightly about the nodules. The frustules seen in girdle view are wedge-shaped, with straight sides and with central nodule visible. There are a number of species, some of which are common. (Pl. I, Fig. 12.)

FAMILY COCCONIDEÆ.—Frustules with valves unlike. Valves broadly oval.

Cocconeis.
Frustules somewhat arched or lens-shaped; in valve view, elliptical or discoidal. Striæ have a general direction transverse to the axis, but the convexity of the frustules gives them the appearance of inclining towards the poles. Upper and lower valves dissimilar, possessing a medial nodule and raphé or pseudo-raphé. (Pl. I, Figs. 13 and 14.)

TRIBE II. PSEUDO-RAPHIDIEÆ.

Possessing a false raphé (simple line or blank space) on one or both valves; with or without nodules. Frustules generally bacillar, sometimes oval or suborbicular, without processes, spines, or awns. Synedra Gaillonii is the typical form.

FAMILY FRAGILARIEÆ.—Frustules adherent, forming a ribbon-like, fan-like, or zigzag filament, or attached by a gelatinous cushion or stipe.

Epithemia.
Frustules cymbiform, symmetrical with respect to the minor axis, with a false raphé and no nodules. Valves marked by lines and pearls approximately at right angles to the major axis, but inclined towards the end of the frustule on the convex side. The frustules in girdle view are seen to be somewhat inflated at the centre. There are several species, differing considerably in the shape of the valves. (Pl. I, Figs. 15 and 16.)

Eunotia.
Frustules elongated, symmetrical with respect to the minor axis. Occurring singly, free-swimming or attached. Valves arcuate, with the convex side undulated. Transversely striated, with two false terminal nodules and no medial line. The frustules

are quadrangular in girdle view. There are but few species, the most common being the E. *tridentula*. (Pl. I, Fig. 17.)

Himantidium.

Sometimes included under Eunotia. The frustules differ from Eunotia by remaining attached after division, forming a band as in Fragilaria; by having the convex side of the valve entire instead of undulate; and by being somewhat bent in girdle view. (Pl. II, Figs. 1 and 2.)

Asterionella.

Frustules long, linear, inflated at the ends. They are united by their extremities into stars or chains, as shown in the girdle view. The typical group is composed of 8 frustules symmetrically and radially arranged. Groups of 4, 6, or 7 are common. When rapidly dividing they may assume a spiral arrangement. The valves are very finely striated, with a straight pseudo-raphé. There is one general species, the A. *formosa*, characterized by having the basal end of the frustules much larger than the free end, and by having on that end a larger surface in contact with the adjoining frustules. There are several varieties, advanced by some authors to the rank of species. The most common is A. *formosa*, var. *gracillima*. (Pl. II, Figs. 3 to 7.)

Synedra.

Frustules elongated, straight or slightly curved. Valves somewhat dilated at the centre and with a medial line or false raphé and occasionally false nodules. They usually have straight and almost, but not quite, parallel-sides. They are finely transversely striated. There are several common species. S. *pulchella* has lanceolate valves, with ends somewhat attenuated. In girdle view they are seen to be attached valve to valve and present the appearance of a long band or a fine-toothed comb. S. *ulna* has a very long rectilinear valve, with conspicuous transverse striæ. There is a false raphé, with a narrow clear space. They are often free-floating. S. *lanceolata* has a long thin valve, swollen at the centre, but tapering to sharp points at the ends. S. *radians* has straight needle-like valves. They are united at the base like Asterionella, but the frustules do not lie in the same plane. (Pl. II, Figs. 8 to 11.)

Fragilaria.
Frustules attached side by side, forming bands as in the case of Synedra pulchella. Valves elongated, straight, with ends lanceolate or slightly rounded. In girdle view the frustules are rectangular and are in contact with each other through their entire length. Valves transversely striated, with a false raphé scarcely visible. There are several common species. (Pl. II, Figs. 12 and 13.)

Diatoma.
Frustules attached by their angles forming zigzag chains, or rarely in bands. In girdle view they are quadrangular. Valves elliptical-lanceolate, with transverse ribs, between which are fine striæ. There is a longitudinal pseudo-raphé. There are two common species,—D. *vulgare* and D. *tenue*. (Pl. III, Figs. 1 to 3.)

Meridion.
Frustules attached valve to valve, forming curved bands seen as fans, circles, or spiral bands. The frustules are wedge-shaped in girdle view, which causes the peculiar shape of the bands. Valves also wedge-shaped, with somewhat rounded ends; furnished with transverse ribs, between which are fine striæ. Pseudo-raphé indistinct. There is one principal species,—M. *circulare*. (Pl. III, Figs. 4 and 5.)

FAMILY TABELLARIEÆ.—Frustules with internal plates, or imperfect septa, often forming a filament.

Tabellaria.
Frustules square or rectangular in girdle view, attached by their corners and forming zigzag chains. In this view they are seen to be marked with longitudinal dividing plates, which extend from the ends not quite to the middle and which terminate in rounded points. The valves are long and thin, and are dilated at the extremities and in the middle. There are fine tranverse striæ and an indistinct pseudo-raphé. The endochrome is usually in rounded lumps. There are two very common species, — T. *fenestrata* and T. *flocculosa*. (Pl. III, Figs. 6 to 9.)

FAMILY SURIRELLEÆ.—Frustules alate or carinate; frequently cuneate.

Nitzschia.
Frustules free, single, elongated, linear, slightly arched, or sigmoidal; with a longitudinal keel and one or more rows of longitudinal points. Valves finely striated, without nodules. There are many species. (Pl. III, Figs. 10 to 12.)

Surirella.
Frustules free, single, furnished with alæ on each side. A transverse section of the frustule shows a double-concave outline. Valves oval or elliptical, with conspicuous transverse tubular striæ, or canaliculi, between which there are sometimes very fine pearled striæ. There is a wide clear space, or pseudo-raphé. The frustules are sometimes cuneate in girdle view. The valves sometimes have a warped surface. There are many common species, most of them of very large size. (Pl. III, Figs. 13 and 14.)

TRIBE III. CRYPTO-RAPHIDIEÆ.

Never possessing a raphé or a false raphé. Frustules generally circular or angular, often provided with teeth, spines, or processes. Stephanodiscus Niagara is the typical form.

FAMILY MELOSIREÆ.—Frustules cylindrical, adhering and forming a stout filament; valves circular, sometimes armed with spines.

Melosira.
Frustules with circular valves and very wide connective bands, attached valve to valve so as to form long cylindrical filaments. In girdle view they are usually rectangular, though sometimes with rounded ends; at the centre there are often conspicuous constrictions. The girdles are often marked with dots. The valves are radially striated, with a clear central space. At the edge there is often a keel or row of projecting points, seen in girdle view. There are several common species. M. *granulata* is the most common free-floating form, and M.

varians, the most common filamentous form. (Pl. III, Figs. 15 to 17.)

FAMILY COSCINODISCEÆ.—Valves circular, generally with radiating cellules, granules, or puncta; sometimes with marginal or intramarginal spines or distinct ribs; without distinct processes.

Cyclotella.

Frustules discoidal, single, occasionally attached valve to valve, but never forming long filaments. Valves circular, finely marked by radial striæ. There is usually an outer ring of radial lines, inside of which there are puncta and fine dots somewhat irregularly arranged. These cannot be seen with low powers. In girdle view the frustules appear rectangular or somewhat sigmoidal, with warped valves, as in C. *operculata*. They are often of very small size. (Pl. III, Figs. 18 and 19.)

Stephanodiscus.

Frustules discoidal, single. Valves circular, with curved surface, with fringe of minute marginal teeth. Striæ fine radial. Frustules rectangular in girdle view, showing projection of middle of valve. Teeth most conspicuous in girdle view. Endochrome conspicuous, in rounded lumps. The frustules are often of considerable size. (Pl. III, Figs. 20 and 21.)

CHAPTER XV.

SCHIZOMYCETES.

THE Schizophyceæ comprise those vegetable organisms in which the chief mode of propagation is that of cell-division. They are either destitute of chlorophyll or contain besides the chlorophyll a coloring substance known as phycocyan or phycochrome, which itself may be a modification of chlorophyll. The cells have a somewhat firm cell-wall, but no nucleus.

The Schizophyceæ may be divided into two classes,—the Schizomycetes and the Cyanophyceæ. The latter contain chlorophyll, but the former do not.

Besides the bacteria, which are not described in this work, there are but four genera belonging to the Schizomycetes that are of interest to the water-analyst. They are so imperfectly understood that no satisfactory classification has been suggested. Some authorities include them among the Fungi.

Leptothrix.
Simple filaments, with indistinct or no articulation, without oscillating movement, and with no sulphur-granules. There are several indistinct species. They are usually colorless. The aquatic forms occur as interwoven masses of long slender filaments, the diameter of which varies from 1 to 3 μ. Lepto-

thrix *ochracea*, observed in driven wells where the water contains much iron, is generally referred to the genus Crenothrix, but its relation to the typical form of Crenothrix is not understood. Very slender forms of Oscillaria are liable to be mistaken for Leptothrix. (Pl. IV, Fig. 1.)

Cladothrix.

Fine filaments resembling those of Leptothrix, colorless, usually indistinctly articulated, straight, undulated, or twisted. There are several stages of development, giving rise to cocci-, vibrio-, spirochætæ-, and filamentous-forms. The special characteristic of the genus is that of false branching, a turning aside of single portions of the filaments followed by subsequent terminal growth. There are several indistinct species. The most important is C. *dichotoma*, which is found in sewage and polluted water. (Pl. IV, Fig. 2.)

Beggiatoa.

Threads indistinctly articulated, colorless, containing numerous dark sulphur-granules. The filaments often have an active oscillating movement. They are usually short and from 1 to 3 μ in diameter. Sometimes abundant in sulphur springs. There are several doubtful species. The most common is B. *alba*. (Pl. IV, Fig. 3.)

Crenothrix.

Filaments cylindrical, transversely divided into cells, surrounded by a gelatinous sheath which becomes yellow or yellowish-brown through deposits of ferric oxide. Multiplication takes place by transverse fission and occasionally by longitudinal fission. Cells also escape from the sheath at the end or side and, by division, form new filaments. Reproduction occurs through spores formed from the cells within the sheath. There is one principal species, C. *Kühniana*. It occurs in single filaments or in brownish tufts or mats, often of considerable thickness. The filaments are 1½ to 4 μ thick, and the sheath is several times the thickness of the filaments. Articulation is distinct. When the iron of the sheath is dissolved by dilute hydrochloric acid the cells appear in side view as distinct rectangles, each one somewhat removed from its neighbor. This appearance is characteristic of Crenothrix.

During growth the cells sometimes push themselves forward in the sheath, leaving the empty sheath behind. The older portion of the sheath is darker colored than the growing points. Crenothrix occurs chiefly in ground-waters rich in organic matter and iron salts. Its growth is favored by darkness. (Pl. IV, Fig. 4.)

CHAPTER XVI.

CYANOPHYCEÆ.

THE plants belonging to the Cyanophyceæ, or Phycochromophyceæ, are characterized by the presence of chlorophyll plus certain coloring substances known as cyanophyll, phycocyanine, phycoxanthine, etc., which are probably modifications of chlorophyll; by the absence of a nucleus and usually of starch-grains; and by extremely simple but imperfectly understood methods of reproduction. The plants are one- or many-celled. By successive division of the cells they are very commonly associated in families that take the form of filaments or of spherical or irregular masses.

The cell-wall is often distinct and sharply defined, but in some cases it is fused with a gelatinous mass in which the cells are embedded. This gelatinous matrix is more common in the terrestrial than in the aquatic species. The cell-contents are usually granular and homogeneous.

The color varies considerably in different species and under different conditions. It is never a chlorophyll green, but ranges from a color approaching that to a blue-green, orange-yellow, brown, red, or violet. The coloring matter known as phycocyanine has a bluish color when viewed by transmitted light, and a reddish color when viewed by reflected light. This phenomenon is often observed in ponds where Cyanophyceæ are abundant. Looking directly at the pond the water may have a reddish-brown color, while a bottle filled with the water and held to the light may present a decidedly bluish-green appearance. This is particularly true when the plants have begun to decay. The phycoxanthine is said to have a yellowish color. The liberation of the gas-bubbles from some species seems to have an effect on the color of the organisms. Anabæna, for example, may have a brownish-green color in a reservoir and a very light blue-green color after it has passed through the pipes of a distribution system, where the pressure has caused the gas to be expelled.

The Cyanophyceæ are usually separated into five or six groups, which are ranked by different writers as *orders*, *families*, or *sections*. The groups are here considered as families belonging to two orders.

ORDER I. CYSTIPHORÆ.

Unicellular plants with spherical, oblong, or cylindrical cells enclosed in a tegument and associated in families, surrounded by a universal tegument or immersed in a generally colorless, mucilaginous substance of varying consistency. Division takes place in one, two, or three directions, the cells after division usually remaining together forming an amorphous thallus. It is probable that most of the forms belonging to this order are but intermediate stages in the life-history of plants higher in the scale of life. There is but one family. It contains about a dozen rather imperfectly defined genera.

FAMILY CHROOCOCCACEÆ.—Thallus mucous or gelatinous, amorphous, enclosing cells and families irregularly disposed.

Chroococcus.
Cells spherical, or more or less angular from compression, solitary or united in small families. Cell-membrane thin or

confluent in a more or less firm jelly. Cell-contents pale bluish-green, rarely yellowish. Propagation by division in three directions. Several species are described. Most of them are terrestrial and not aquatic. The most common aquatic species are C. *turgidus*, the cells of which are from 10 to 25 μ in diameter, and C. *cohærens*, the cells of which are from 3 to 6 μ in diameter. (Pl. IV, Fig. 5.)

Glœocapsa.
Cells spherical, single or in groups; each cell surrounded by a vesculiform tegument and groups of cells surrounded by an additional tegument. Cell-membrane thick, lamellated, and sometimes colored. Division in three directions. Cell-contents bluish-green, brownish, or reddish. There are many described species, based on slight distinctions and variations in size and color. Glœocapsa found in water usually has smaller cells and a more distinct tegument than Chroococcus. Comparatively few species are aquatic. (Pl. IV, Fig. 6.)

Aphanocapsa.
Cells spherical, with a thick, soft, colorless tegument, confluent in a homogeneous mucous stratum which is sometimes of a brownish color. Cell-contents bluish-green, brownish, etc. The cells divide alternately in three directions. There are several species. The cells vary in size from 3 to 6 μ. (Pl. IV, Fig. 7.)

Microcystis.
Cells spherical, numerous, densely aggregated, enclosed in a very thin, globose mother-vesicle, forming solid families, singly or several surrounded by a universal tegument. Cell-contents æruginous to yellowish-brown. The cells divide alternately in three directions. This genus represents a condition of frequent occurrence in the process of development of higher forms. There are several indistinct species common in water. The cells vary in size from 4 to 7 μ in diameter and the colonies from 10 to 100 μ. (Pl. IV, Fig. 8.)

Clathrocystis.
Cells very numerous, small, spherical or oval, æruginous, embedded in a colorless matrix. Multiplication by division of the cells within the thallus. The thallus is at first solid, then

becomes saccate and clathrate (perforated); broken fragments are irregularly lobed. There is but one species,—C. *æruginosa*. The cells are from 2 to 4 μ in diameter and the thallus from 25 μ to 5 mm. This species is widely distributed. (Pl. IV, Fig. 9.)

Cœlosphærium.

Cells numerous, minute, globose or subglobose, geminate, quaternate, or scattered, immersed in a mucous stratum. Cell-contents æruginous, granulose. The thallus is globose, vesicular, hollow, the cells being found only on the outer surface. Multiplication takes place by division of the cells on the surface and by the escape and further development of certain peripheral cells. There is one common species, C. *Kuetzingianum*. The cells are from 2 to 5 μ in diameter and the thallus from 50 to 500 μ. (Pl. IV, Fig. 10.)

Merismopedia.

Cells globose or oblong, æruginous or brownish, with confluent teguments. Division in two directions. The thallus is tabular, quadrate, free-swimming, the cells being arranged in groups of 4, 8, 16, 32, 64, 128, etc. There are several indistinct species. The diameter of the cells varies from 3 to 7 μ. (Pl. IV, Fig. 11.)

Glœothece.

Similar to Glœocapsa, but with oblong or cylindrical, instead of spherical cells. Terrestrial rather than aquatic.

Aphanothece.

Similar to Aphanocapsa, but with oblong instead of spherical cells.

Tetrapedia.

Cells compressed, quadrangular, equilateral, subdivided into quadrate or cuneate segments or rounded lobes, either by deep incisions or wide angular sinuses. This genus is of doubtful value.

ORDER II. NEMATOGENÆ.

Multicellular plants, the cells of which dividing in one direction, form filaments, often enclosed in a tubular sheath.

The filaments (trichomes) may be either simple or branched. There are five families.

FAMILY NOSTOCACEÆ.—Plants composed of rounded cells loosely united into filaments, or trichomes, and sometimes embedded in jelly. The filaments do not branch and never terminate in a hair-point. They sometimes forms large masses. There are three kinds of cells—ordinary vegetative cells, joints, or articles; heterocysts; and spores. The ordinary cells are spherical, elongated, or compressed. The cell-contents are bluish-green or brownish, and are usually granular. The heterocysts are cells found at intervals in the filaments. They are spherical, elliptical, or elongated, and are usually somewhat larger than the vegetative cells. Their cell-contents are generally clear or very finely granular, and usually of a light bluish-green color. The cell-wall is sharply defined, and there are two polar lumps of gelatinous material that cause them to adhere to the adjoining cells. The function of the heterocysts is unknown, but they are thought to be in some way connected with the process of reproduction. The spores are usually much larger than the vegetative cells. They are spherical, elliptical, or cylindrical. Their cell-contents are usually very granular and dark-colored. They seem to be more highly differentiated than the contents of the vegetative cells. The spores are heavy, and will sink in water when freed from the filaments. Multiplication takes place by division of the vegetative cells, by means of the spores, and by means of hormogons, or parts of the internal trichomes which separate from the filaments and form new plants. The character and position of the heterocysts and spores form the chief basis for the division of the Nostocaceæ into genera. The classification is very indefinite.

Nostoc.

Cells globose or elliptical; heterocysts usually globose and somewhat larger than the vegetative cells; spores oval and but little larger than the heterocysts. Spores and heterocysts are both intercalated in the filaments, rarely terminal. The filaments are enclosed in a gelatinous envelope, and are flexuously curved and irregularly interwoven. They often form gelatinous fronds or thalli surrounded by a firm membrane. The thalli vary in diameter and are sometimes of great size. There are many species, both terrestrial and semi-aquatic. The species are not well defined, and many of them are intermediate stages in the life-history of higher forms. The true Nostoc is seldom found in drinking-water. (Pl. IV, Fig. 12.)

Anabæna.

Vegetative cells spherical, elliptical, or compressed in a quadrate form. Heterocysts much larger than the vegetative cells, subspherical, elliptical, or barrel-shaped, of a pale yellowish-green color, and intercalated in the filament. Spores globose or oblong-cylindrical, equal to or somewhat larger than the heterocysts, rarely smaller, never adjacent to the heterocyst. The filaments are moniliform; are without sheaths; are straight, curved, circinate, or intertwined; have a bluish-green or brownish color; and are often free-floating. There are several important but imperfectly defined species. The most common species are *A. flos-aquæ* and *A. circinalis*. The vegetative cells of the former are from 5 to 7 μ in diameter; those of the latter are from 8 to 12 μ. (Pl. IV, Figs. 13 and 14.)

Sphærozyga.

Vegetative cells spherical, elliptical, or transversely compressed; of a bluish-green or brownish color. Heterocysts spherical or oval, intercalated, binary or solitary, only slightly larger than the vegetative cells. Spores on each side of and adjacent to the heterocysts, cylindrical, with rounded ends, considerably larger than the heterocysts. The filaments are moniliform; are sheathless or covered with a mucilaginous coating, occasionally agglutinated in a gelatinous stratum. There are several species, terrestrial and aquatic. The genus is very similar to Anabæna. (Pl. V, Fig. 1.)

Cylindrospermum.

Vegetative cells globose, elliptical, or compressed, homogeneous or granular. Heterocysts terminal, spherical, or oval, but little larger than the cells. Spores adjacent to the heterocysts, oval or cylindrical, much larger than the cells. The filaments are moniliform, sheathless, and sometimes taper slightly. There are few species, and these resemble some forms of Anabæna and Sphærozyga. (Pl. V, Fig. 2.)

Aphanizomenon.
Vegetative cells cylindrical, closely connected, granular, and with little color. Heterocysts rare, intercalated, oval, but little larger in diameter than the cells. Spores very rare, intercalated, not adjacent to heterocysts, cylindrical, with rounded ends, sometimes of dark olive color. The filaments are cylindrical, slightly tapering, and densely agglutinated in fascicles, occasionally free. The fascicles are often of considerable size. Diameter of filaments 4 to 6 μ. This genus is sometimes mistaken for Oscillaria or Anabæna. (Pl. V, Fig. 3.)

FAMILY OSCILLARIEÆ (LYNGBYÆ).—Filaments without heterocysts or spores, with or without sheath, not terminating in a hair-point, single or associated in bundles enclosed in a common sheath. The division of the filaments into cylindrical cells is indistinct. Multiplication is said to take place by hormogons, i.e. parts of the trichomes which separate from the rest of the filament.

Oscillaria.
Cells shortly cylindrical, disc-shaped in end-view, closely united into a simple, branchless, sheathless filament. The filaments are straight or somewhat curved, occasionally fasciculate, and have rounded ends. The color is bright bluish-green, steel-blue, etc. The filaments when in active vegetative state possess characteristic spontaneous oscillating movements. There is a large number of species, that vary in diameter from 1 to 50 μ, and have cells differing in shape and in color. There are but few free-floating forms. (Pl. V, Fig. 4.)

Lyngbya.
Filaments enclosed singly in a sheath, branchless, but with occasional appearance of branching during multiplication, sometimes combined to form a membranaceous stratum. Cells united into short trichomes, with rounded ends, not continuous in the sheath, but separated by clear spaces. Cell-contents blue-green, granular. Sheaths pellucid, hyaline. Propagation is said to take place by hormogons and by gonidia. There are many species, terrestrial and aquatic. (Pl. V, Fig. 5.)

Microcoleus.
>Filaments rigid, articulate, crowded together in bundles, enclosed in a common mucous sheath, either open or closed at the apex. Sheath ample, colorless, rarely indistinct. Several species, chiefly terrestrial. (Pl. V, Fig. 6.)

FAMILY SCYTONEMEÆ. —Filaments with lateral ramifications (false branching) in which some of the cells change into heterocysts; enclosed in a sheath. The cells divide transversely. The ramifications are produced by the deviation of the trichome and emergence through the sheath. The branches do not have a hair-point. There are several genera.

Scytonema.
>Sheath enclosing a single trichome, composed of subspherical or subcylindrical cells, with scattered heterocysts. Color bluish- or yellowish-green. Ramification takes place by a folding of the trichomes, followed by rupture of the sheath and the emergence of one or two portions of the folded trichome at right angles to the original filament. These branched filaments produce interwoven mats. Multiplication is said to take place by microgonidia. There are many species, terrestrial and aquatic. The plant is not found free-floating. (Pl. V, Fig. 7.)

FAMILY SIROSIPHONEÆ. —Trichomes enclosed in an ample sheath, profusely branched. Branches are formed by longitudinal division of certain cells so as to form two sister cells, the inferior of which remains a part of the trichome, while the other, by repeated division, grows into a branch. The filaments often contain 3, 4, or more series of cells. Propagation is said to take place by means of microgonidia.

Sirosiphon.
>Cells one-, two-, or many-seriate, in consequence of their lateral division or multiplication. The cells have a distinct membrane and the sheaths are large. The plant is never found free-floating. (Pl. V, Fig. 8.)

FAMILY RIVULARIEÆ. — Filaments free or agglutinated into a definite thallus, terminating at the apex in a hair-like extremity. Heterocysts usually basal. Trichomes articulated like Oscillaria, parallel or radially disposed. Spores, when present, cylindrical, generally adjacent to the basal heterocyst.

Rivularia.
Filaments radial, agglutinated by a firm mucilage, and forming well-defined hemispherical or bladdery forms. Heterocysts basal. No spores formed. Ramifications produced by transverse division of the trichomes. Color greenish to brownish. Sheaths usually distinct. Several species, terrestrial and aquatic. Occasionally found free-floating. (Pl. V, Fig. 9.)

CHAPTER XVII.

CHLOROPHYCEÆ.

THE Algæ are flowerless plants of simple cellular structure, without mycelia, roots, stems, or leaves. The functions of the plants are centred in the individual cells, and only to a limited extent is there any "division of labor" among the cells.

It is difficult to define the word "Algæ," because it is used differently by different writers. In the broad sense it includes all of the thallophytes which contain chlorophyll, i.e. the Diatomaceæ, Cyanophyceæ, Chlorophyceæ, Phæophyceæ, and Rodophyceæ. This is the older meaning of the term. It is used in contradistinction to the Fungi, which contain no chlorophyll. In the narrower sense it includes only the Chlorophyceæ, Phæophyceæ, and Rodophyceæ. This is the later and better use of the word. The Phæophyceæ and Rodophyceæ are almost entirely marine forms, so that, as far as fresh-water forms are concerned, the word algæ is almost synonymous with Chlorophyceæ. In popular speech, however, the Cyanophyceæ are frequently spoken of as the "blue-green algæ," and the diatoms have sometimes been called the "brown algæ."

The plants belonging to the Chlorophyceæ are characterized by the presence of true chlorophyll, a nucleus, starch-grains, and often by a cell-wall made of cellulose. They are "algæ" in the strictest sense of the term. They cover a great range of complexity. Some of them are minute, unicellular forms scarcely distinguishable from the Cyanophyceæ; others resemble the Protozoa; while others are large, branching, multicellular forms doubtfully included among the algæ, and very similar to plants much higher in the scale of life. Most of them are aquatic, but a few are terrestrial. Their color is almost always a bright chlorophyll green, but occasionally it is yellowish-brown or even a bright red. The Chlorophyceæ increase by the ordinary processes of cell-division observed in the higher forms of plant life. The cells may separate after division, or they may remain associated in colonies or in simple or branching filaments. Reproduction takes place either asexually, i.e. without the aid of fecundation, or sexually. There is but one general method of asexual reproduction, namely, the formation within the cell of spores, which become scattered and give rise to new cells. There are three general types of sexual reproduction. The simplest is the formation in the cells of zoöspores, which become liberated and ultimately copulate with other zoöspores. Two of these zoöspores become attached by their ciliated ends, their contents become fused, and a zygospore results. After a period of rest the zygospore may develop into a new plant, or may break up into other spores. The second type of sexual reproduction is known as conjugation. Two cells come in contact, and by means of openings in the cell-walls their contents become fused. A zygospore (sometimes two) is formed, which, after a period of rest, gives rise to new plants. The highest form of sexual reproduction takes place by the formation of a rather large female oospore, which becomes fertilized by small male cells or spermatozoids. This mode of reproduction is analogous to that observed in the higher plants. Many of the Chlorophyceæ exhibit the phenomenon of "alternation of generations," by which is meant the continued propagation of the plants by asexual processes with occasional intervention of the sexual processes.

ORDER I. PROTOCOCCOIDEÆ.

Unicellular plants. Cells single or associated in families; tegument involute or naked; no branching or terminal vegetation. This order includes many of the free-floating green algæ that are found in water.

FAMILY PALMELLACEÆ.—Cells solitary or in families, often embedded in a jelly and forming an amorphous stratum. Multiplication by cell-division. Reproduction asexual, by active gonidia.

Glœocystis.

Cells globose or oblong, single or in globose families of 2–4–8 cells. Common and individual lamellose gelatinous integuments. Division in alternate directions. Reproduction by zoogonidia. There are several species. The size of the cells varies from 2 to 12 μ in diameter and the colonies from 10 to 100 μ. Color green, sometimes reddish. Gelatinous tegument colorless or ochraceous. Usually fixed, sometimes free-floating. (Pl. V, Fig. 10.)

Palmella.

Cells globose, oval, or oblong, surrounded by a thick confluent tegument; forming an amorphous thallus. Multiplication by alternate division of the cells in all directions. An uncertain genus. Several species, usually fixed. Size of cells varies from 1 to 15 μ. Thallus often large. Color generally green. (Pl. V, Fig. 11.)

Tetraspora.

Cells spherical or angular, with thick teguments confluent into a homogeneous mucous; forming a sac-like thallus, sometimes of large size. The cells divide in two directions and are seen normally in groups of four. The thalli are usually fixed, but the quartettes of cells are sometimes free-floating.

Several species, all green. Cells from 3 μ to 12 μ in diameter. (Pl. V, Fig. 12.)

Botryococcus.
Cells generally oval, with a thin confluent tegument, densely packed, forming a botryoid, irregularly lobed thallus. One species, green, free-floating, with cells 10 μ in diameter. (Pl. VI, Fig. 1.)

Raphidium.
Cells fusiform or cylindrical, straight or curved, pointed ends, occurring singly, in pairs, or in fascicles. Cell-membrane thin, smooth. Cell-contents green, granular, with transparent vacuole. Division of cells in one direction. There are several species, with numerous varieties. Two species, R. *polymorphum* and R. *convolutum*, are common free-floating forms. The latter is sometimes known by the name Selenastrum. Pl. VI, Fig. 2.)

Dictyosphærium.
Cells elliptical or kidney-shaped, with thick mucous investment, more or less confluent, arranged in globose, hollow families. The cells are connected by delicate threads radiating from the centre of the colony and attached to the concave side of the cells. The threads branch dichotomously. Division in all directions. Two or three species. The most important species is D. *reniforme*. Color green, and cells 6–10 × 10–20 μ. (Pl. VI, Fig. 3.)

Nephrocytium.
Cells oblong, kidney-shaped, with ample tegument, arranged in free-swimming colonies of 2–4–8–16 cells. Two species. Green. Cells 5 × 15 to 15 × 45 μ. (Pl. VI, Fig. 4.)

Dimorphococcus.
Cells in groups of four on short branches, the two intermediate contiguous cells oblique, obtuse-ovate; the two lateral, opposite and separate from each other, lunate. In colonies with cells connected by threads radially arranged and unbranched. One free-floating species. Color green. Cells 5 to 10 μ in diameter. (Pl. VI, Fig. 5.)

FAMILY PROTOCOCCACEÆ.—Cells solitary or forming more or less perfect cœnobia. Propagation by asexual zoospores or by copulation of zoogonidia. In general there is no vegetative cell-division.

Protococcus.

Cells spherical, single or in irregular clusters. Cell-membrane thin, hyaline. Cell-contents green, sometimes reddish. There is but one species, P. *viridis*, with many varieties. Diameter of cells varies from 3 to 50 μ. They are both aquatic and aerial. Some of the aquatic forms have a gelatinous tegument and are called Chlorococcus by some writers. The distinction is a difficult one to make. (Pl. VI, Fig. 6.)

Polyedrium.

Cells single, segregate, free-swimming, compressed, 3–4–8-angled. Angles sometimes radially elongated, entire or bifid, rounded at the ends. Cell-membrane thin, even. Cell-contents green, granular, sometimes with oil-globules. Propagation by gonidia. There are several species. One of the most common is P. *longispinum*. (Pl. VI, Fig. 7.)

Scenedesmus.

Cells elliptical, oblong, or cylindrical, with equal or unequal ends, often produced into a spine-like horn; usually laterally united, forming cœnobia. Cell-contents green. Propagation by segmentation of cell-contents into brood families, set free by rupture of the maternal cell-membrane. There are several common species, S. *caudatus*, with several varieties, S. *obtusus*, and S. *dimorphous*. The cells are usually 2 or 3 μ in diameter and from 8 to 25 μ long. (Pl. VI, Fig. 8.)

Hydrodictyon.

Cells oblong-cylindrical, united at the ends into a reticulated, saccate cœnobium. Cell-contents green. Propagation by macrogonidia which join themselves into a cœnobium within the mother cell, and by ciliated microgonidia which copulate and form a resting-spore. One species, H. *utriculatum*. Aquatic and attached. (Pl. VI, Fig. 9.)

Ophiocytium.

Cells cylindrical, elongated, curved, or circinate, one end and occasionally both ends attenuated. Cell-contents green. Propagation by zoogonidia. There are several species. The most common is O. *cochleare*, the cells of which are from

5 to 8 μ in diameter and of various lengths. (Pl. VI, Fig. 10.)

Pediastrum.

Cells united into a plane, discoid or stellate, free-swimming cœnobium, which is continuous, or with the cells interrupted in a perforate or clathrate manner. The central cells are polygonal and entire; those of the periphery entire, bi-lobed, with lobes sometimes pointed. Cell-contents green, granular. Propagation by macrogonidia formed within the cells, which after their escape divide, arrange themselves in a single layer, and reproduce the form of the mother plant. There are several species. The most common are P. *Boryanum* and P. *simplex*. (Pl. VI, Fig. 11.)

Sorastrum.

Cells wedge-shaped, compressed, sinuate, emarginate, or bifid at the apex; radially disposed, forming a globose, solid, free-swimming cœnobium. There is but one species, S. *spinulosum*. The cells are spined. They vary in size from 12 to 20 μ. The cœnobia vary in diameter from 25 to 75 μ. (Pl. VI, Fig. 12.)

Cœlastrum.

Cells globose, or polygonal from pressure, forming a globose, hollow cœnobium, reticulately pierced. The cells are arranged in a single layer, sometimes joined by radial gelatinous cords. Cell-contents green. Propagation by macrospores. There are several species. The most common is C. *microporum*, which has 8-16-32 cells, and the diameter of which varies from 40 to 100 μ. (Pl. VII, Fig. 1.)

Staurogenia.

Cells oblong-oval, subquadrate, or rhomboidal, arranged in groups of 4-8-16, forming a cubical cœnobium, hollow within. Cell-contents green. Propagation by quiescent gonidia. (Pl. VII, Fig. 2.)

ORDER II. VOLVOCINIEÆ.

Unicellular plants occurring as mobile, globose, subglobose, or flattened quadrangular cœnobia composed of bi-ciliated green cells which are more or less spherical or compressed. The cœnobia as a whole are motile because of the ciliated cells, and hence are free-floating. The cœnobium sometimes has an ample hyaline tegument. Cell-contents green. Propagation sexual or asexual. Asexual propagation takes place by subdivision of the larger vegetative cells into new families, which separate from the mother cell when sufficiently developed. Sexual propagation takes place by means of female spore-cells, or oospores, developed from the vegetative cells, which are fertilized by antheridia developed from other vegetative cells. The antheridia, after escaping from the cell in which they are formed, perforate the membrane of the oogonia, after which the oospore goes into a resting state to germinate later. This order is frequently referred by zoologists to the Protozoa.

FAMILY VOLVOCACEÆ.—Characteristics the same as for the order.

Volvox.

Large cœnobium, continually rotating and moving, looking like a hollow globe composed of very numerous cells (several thousand) arranged on the periphery at regular distances, connected by a matrical gelatin which has the appearance of a membrane in which the cells are embedded. Cells globose, bearing two cilia that extend beyond the gelatinous envelope. By the waving of these cilia the colony is kept in motion. Cell-contents green; starch-granules and often a red pigment-spot present. With a high power the cells are seen to be connected to each other in a hexagonal manner by fine threads. Propagation sexual and asexual, as described under the order. The oospores and antheridia are enclosed in flask-like cells extending inward. The spermatozoids are spindle-shaped and furnished with two cilia. The resting-spores usually produce eight zoogonidia. Asexual propagation takes place by division of the larger and darker flask-like cells. These, usually eight in number, develop young volvoxes in the mother cells. They are very conspicuous. The mother cell splits along well-defined lines and the young forms are set free. There is practically but one species, *V. globator*. The cœnobia are often one millimeter in diameter. (Pl. VII, Fig. 3.)

Eudorina.
Cœnobium oval or spherical, involved in a gelatinous mucilaginous tegument, composed of 16–32 cells arranged around the colorless sphere at equal distances. The cœnobium is often seen moving with a rolling motion. Cells globose, with two protruding cilia. Cell-contents green, sometimes with a red pigment-spot. Asexual propagation takes place by the division of the cells into 16–32 parts, each of which produces a new cœnobium. Sexual propagation as described for the order. Usually four of the thirty-two cells produce antheridia, the others oogonia. The spermatozoids are pear-shaped and are bi-ciliated. There are but two species, *E. elegans*, and *E. stagnale*. The cells vary from 5 to 25 μ, and the cœnobia from 25 to 150 μ, in diameter. (Pl. VII, Fig. 4.)

Pandorina.
Cœnobium globose, invested by a broad, colorless, gelatinous tegument, composed of 8 to 64 cells crowded together or aggregated in a botryoidal manner. (In this respect it differs from Eudorina.) Cells green, globose or polygonal from compression, bi-ciliated, occasionally with a red pigment-spot. Sexual propagation takes place by the conjugation of zoospores produced in the cells of the cœnobium, which after union give rise to resting-spores. Asexual propagation takes place by cell-division. There is but one species, *P. morum*. The cœnobium is about 200 μ in diameter and the cells from 10 to 15 μ. (Pl. VII, Fig. 5.)

Gonium.
Cœnobium quadrangular, tabular, with rounded angles, formed from a single flat stratum of cells, girt by a broad, hyaline, plane-convex tegument. Cells 16 (4 central and 12 peripheral), polygonal, connected by produced angles, and furnished with two cilia. Cell-contents green. Asexual propagation by division. Sexual propagation unknown. There is

but one species, G. *pectorale*. The genus is an uncertain one. (Pl. VII, Fig. 6.)

ORDER III. CONJUGATE.

Unicellular or multicellular plants. The multicellular forms have no terminal vegetation and are destitute of true branches. The chlorophyll masses are arranged in plates, bands, or stellate masses. Starch-grains are abundant. Multiplication by division in one direction. Reproduction by zygospores resulting from copulation and conjugation of two cells, or by azygospores formed without copulation. There are two families that are very different in their general characteristics, but that agree in their mode of reproduction.

FAMILY DESMIDIEÆ.—The Desmidieæ, or Desmids, form a large, well-defined group of unicellular algæ. They are characterized by two peculiar features,—by an apparent division of the cell into two symmetrical halves, and by the presence of projections from the surface, either inconspicuous or prolonged into spines. The cells are of various sizes and forms, often curious or ornamental, single or joined together forming a filament. The transverse constriction is sometimes deep, sometimes slight, and occasionally absent. The cell-wall is firm, almost horny. Some writers have imagined that it was slightly silicified. The cell is surrounded by a mucous covering and sometimes by a layer of gelatin. The cell-contents are green and granular. Starch-grains are numerous. At the ends of some of the cells there are clear spaces in which are seen granules that occasionally have a vibratory movement. Cyclosis, or a circulation of granules in the watery fluid next the cell-wall, may be observed in some species. Some species of desmids exhibit voluntary movements of the entire cell. Closterium, for example, shows certain oscillations and backward and forward gliding movements, supposed to be due to the secretion of threads of mucous. Multiplication takes place by cell-division and by conjugation. In the first case the two halves of the cell stretch apart and become separated by a transverse partition; new halves ultimately form on each of the original halves, so that two symmetrical cells result. These afterwards separate. (See Pl. VIII, Fig. A.) Sexual propagation by conjugation takes place as follows: Two cells approach and each sends out a tube from its centre. These tubes meet, swell hemispherically, and, by the disappearance of the separating wall, become united into a rounded zygospore with a thick tegument and sometimes with bristling projections. This zygospore, after a period of rest, loses its contents through a rent in the wall, and a new cell is formed which ultimately becomes constricted and assumes the shape of the parent cell. (See Pl. VIII, Figs. B to F.)

Some of the common genera are described below. The enormous number of species makes a detailed analysis impracticable.

Penium.
Cells straight, cylindrical or fusiform, not incised nor constricted in the middle; ends rounded. Chlorophyll lamina axillary; containing starch-granules. Cell-membrane smooth, finely granulated, or longitudinally striated. Individuals free-swimming or associated in gelatinous masses. (Pl. VII, Fig. 7.)

Closterium.
Cells simple, elongated, lunate or crescent-shaped, entire, not constricted at the centre. Cell-wall thin, smooth or somewhat striated. The chlorophyllaceous masses are generally arranged in longitudinal laminæ, interrupted in the middle by a pale transverse band. At each end there is a clear, colorless, or yellowish vacuole in which minute "dancing granules" may be seen. (Pl. VII, Figs. 8 to 10.)

Docidium.
Cells straight, cylindrical or fusiform, elongated, constricted at the middle. The semi-cells are somewhat inflated at the base and are often separated by a suture. Ends rounded, truncated or divided. Transverse section circular. The chlorophyllaceous cytioplasm has a parietal or axillary arrangement. Terminal vacuoles with "dancing granules" are observed in some species. (Pl. VII, Fig. 11.)

Cosmarium.
Cells oblong, cylindrical, elliptical, or orbicular, with margins smooth, dentate, or crenate; deeply constricted; ends rounded or truncate and entire; end view oblong or oval. Chloro-

phyll masses parietal or concentrated in the centre of the semi-cells. Cell-walls smooth, punctate, warty, or rarely spinous. The zygospore is spherical, tuberculated or spinous. (Pl. VII, Fig. 12, and Pl. VIII, Figs. A to F.)

Tetmemorus.
Cells cylindrical or fusiform, slightly constricted in the middle, narrowly incised at each end, but otherwise entire. Cell-wall punctate or granulate. (Pl. VII, Fig. 13.)

Xanthidium.
Cells single or geminately concatenate, inflated, very deeply constricted; semi-cells compressed, entire, spinous, protruding in the centre as a rounded, truncate, or denticulate tubercle. Cell-wall firm, armed with simple or divided spines. The zygospores are globose, smooth or spinous. (Pl. VIII, Figs. 1 and 2.)

Arthrodesmus.
Cells simple, compressed, deeply constricted; semi-cells broader than long, with a single spine on each side, but otherwise smooth and entire. (Pl. VIII, Fig. 3.)

Euastrum.
Cells oblong or elliptical, deeply constricted; semi-cells emarginate and usually incised at their ends; sides symmetrically sinuate or lobed, provided with circular inflated protuberances; viewed from the vertex, elliptical. The zygospores are spherical, tuberculose or spinous. (Pl. VIII, Fig. 4.)

Micrasterias.
Cells simple, lenticular, deeply constricted; viewed from front, orbicular or broadly elliptical; viewed from the vertex, fusiform, with acute ends; semi-cells three- or five-lobed; lateral lobes entire or incised; end lobes sinuate or emarginate and sometimes with angles bifid or produced. (Pl. VIII, Fig. 5.)

Staurastrum.
Cells somewhat similar to those of Cosmarium in front view, but angular in end view; angles obtuse, acute, or drawn out into horn-like processes. Cell-wall smooth, punctate or granular, hairy, spinulose, or extended into arms or hair-like processes. Chlorophyll masses concentrated at the centre of the

semi-cells, with radiating margins. The zygospores are spined. (Pl. VIII, Figs. 6 and 7.)

Hyalotheca.
Cells short, cylindrical, usually with a slight obtuse constriction in the middle; circular in end view. The cells are closely united into long filaments, enclosed in an ample, colorless mucous sheath. The chlorophyll is concentrated in a mass which, in end view, has a radiate appearance. (Pl. IX, Fig. 1.)

Desmidium.
Cells oblong-tabulate, somewhat incised; in end view, triangular or quadrangular; united into somewhat fragile filaments and surrounded by a colorless mucous sheath. Chlorophyll masses in each semi-cell concentrated and radiate to the angles. Zygospores smooth, globose or oblong. (Pl. IX, Fig. 2.)

Sphærozosma.
Cells bi-lobed, elliptical, or compressed, deeply incised, forming filaments which are almost moniliform or pinnatifid, surrounded by a colorless or mucous sheath. Chlorophyll mass concentrated, somewhat radiate. (Pl. IX, Fig. 3.)

FAMILY ZYGNEMACEÆ.—Multicellular plants, composed of cylindrical cells joined into filaments and forming an articulated simple thread. Cell-wall lamellose. Chlorophyll arranged as twin stellate nuclei, as axillary laminæ, or as spiral bands. Starch-grains, etc., conspicuous. Propagation by zygospores resulting from copulation, which takes place by the union of two filaments. The filaments come into proximity, the cells put out short processes, which unite, forming tubular passages between pairs of cells. Through these connecting tubes the cell-contents of one cell passes into and unites with the cell-contents of another. This results in the formation of a zygospore often clothed with a triple membrane. Copulation is said to be scalariform when opposite cells of two filaments unite by ladder-like tubes, geniculate when the cells become bent and unite at the angles, and lateral when the process takes place between two adjoining cells of the same filament. The family is sometimes divided into two sections, the *Zygneminæ* and *Mesocarpinæ*. In the second section the spore formed is not a true zygospore. It is formed by a flowing together of only a part of the cell-contents. The zygospores germinate by putting forth a single germ, which elongates by transverse division into a filament.

Spirogyra.
Cells cylindrical, sometimes replicate, or folded in at the ends. Chlorophyll arranged in one or several parietal spiral bands winding to the right. Copulation scalariform, sometimes lateral. Copulating cells often shorter than sterile ones and more or less swollen. Zygospores always within the wall of one of the united cells. There are very many species, differing in size of cells, number and arrangement of spirals, replication at the end of cells, character of the zygospore, etc. (Pl. IX, Figs. 4 and 5.)

Zygnema.
Cells with two-axil, many-rayed chlorophyll bodies near the central cell-nucleus, containing one or more starch-granules. Copulation scalariform or lateral. Zygospore in one of the united cells. (Pl. IX, Fig. 6.)

Zygogonium.
Like Zygnema, except that the zygospores are located in the connecting tube between the united cells.

ORDER IV. SIPHONEÆ.

Unicellular plants when in the vegetative state; cells tubular or utricle-shaped, often branched. Cell-contents green, granular. Propagation by sexual fertilization, asexual zoospores, or by microgonidia.

FAMILY VAUCHERIACEÆ.—Plants consisting of elongated, robust tubular filaments, more or less branched, growing in tufts. Chlorophyll granules are evenly distributed on the inside walls of the cells, and starch-grains and oil-globules are conspicuous. Sexual propagation takes place by means of oospores fertilized by spermatozoids. The oogonia are lateral, sessile, or borne on a simple pedicel; the antheridia usually develop on the same filament. Asexual propagation takes place by means of zoospores produced in a terminal sporangium. The zoospores are ciliated, but go through a resting period before germinating. Propagation also takes place by means of microgonidia produced in the vegetative cells.

Vaucheria.
> The characteristics are described under the family. There are many species, aquatic and terrestrial. (Pl. IX, Fig. 7.)

ORDER V. CONFERVOIDEÆ (NEMATOPHYCEÆ).

Multicellular plants consisting of simple or branched filaments forming articulated threads or membranaceous thalli. Vegetation terminal, sometimes lateral. Propagation by oospores fertilized by spermatozoids, or by copulation of zoogonidia. In many of the genera the method of propagation is not well known. The order contains a great variety of forms, and various methods of classification have been adopted by different writers. There are but few genera that interest the water-analyst

FAMILY CONFERVACEÆ.—Plants consisting of simple or branched filaments, with terminal vegetation, composed of elongated, cylindrical cells, rarely abbreviated or swollen. Cell-membrane sometimes lamellose. Vegetation by division in one direction. Propagation by zoospores.

Conferva.
> Articulate threads simple; cells cylindrical, sometimes swollen; chlorophyll homogeneous. Vegetation by division. Propagation by zoogonidia. There are many common species, varying greatly in diameter of filaments. Many vegetative filaments of other plants are liable to be mistaken for Conferva. The characteristics of the genus are somewhat vague. (Pl. IX, Fig. 8.)

Cladophora.
> Articulate threads very much branched, the branched cells being much thinner than the primary cells. Cell-membrane thick, lamellose. Cells cylindrical, somewhat swollen. Cell-

contents green, containing many starch-granules. Propagation by zoogonidia, which develop in large numbers. (Pl. IX, Fig. 9.)

FAMILY ŒDOGONIACEÆ. — Filaments articulated, simple or branched. Cells cylindrical, terminal cells sometimes setiform. Propagation by asexual zoospores or by oospores sexually fertilized. Plants monœcious or diœcious; when diœcious the male plants are either dwarf, i.e. produced from short cells of the female plants, or elongated and independent. There are two genera, Œdogonium and Bulbochæte, each with many species.

FAMILY ULOTRICHEÆ.—Filaments shortly articulate, simple, free, sometimes laterally connate in bands. Cell-membrane thick and lamellose. Cell-contents at first effused, after division transmuted into gonidia. Propagation by ciliated macrospores which do not copulate, or by microzoospores which do or do not copulate.

Ulothrix.
Filaments simple, articulate. Articulations usually shorter than their diameter. Cell-membrane thin. Cell-contents green, effused or parietal, enclosing amylaceous granules. Propagation by macro- and micro-zoospores. Several common species. (Pl. IX, Fig. 10.)

FAMILY CHÆTOPHORACEÆ.—Filaments articulate, dichotomously or fasciculately branched, accumulated in tufts in a gelatinous mucus, or constituting a filamentose or foliaceous thallus. Propagation by oospores sexually fertilized, or by zoogonidia. Monœcious or diœcious.

Stigeoclonium.
Filaments articulate, with simple scattered branches. Branches similar to the stems, attenuated into a colorless bristle. Cell-membrane thin, hyaline. Cell-contents green, with chlorophyll arranged in transverse bands. Propagation by oospores or zoogonidia. (Pl. X, Fig. 1.)

Draparnaldia.
Filaments articulate, much branched; the main stem comparatively thick, composed of large, mostly hyaline cells, with broad, transverse chlorophyll bands. Many branches and sub-

branches, alternate or opposite. The terminal cells are empty, hyaline, and often elongated into a bristle. The branch cells only are fertile. The plant is enveloped in a gelatinous covering. Propagation by resting-spores or zoogonidia. There are few species. (Pl. X, Fig. 2.)

Chætophora.
Filaments articulate, with primary branches radiately disposed, and secondary branches shortly articulate, and attenuated into a bristle, the whole involved in a gelatinous mass. Propagation by zoospores. (Pl. X, Fig. 3.)

ORDER VI. CHARACEÆ.

The Characeæ are plants which occupy an intermediate position between the algæ and the higher cryptogams. Each plant consists of an assemblage of long tubular cells, having a distinct central axis, with whorls of branches projecting at regular intervals at points called "nodes." The branches are sometimes spoken of as leaves, but they are quite similar to the stem. At the lower end of the stem some of the branches (rhizoids) are root-like and serve to give attachment and stability to the plant. Reproduction takes place by a peculiar sexual process. Oospheres or archegones form at the base of the branches and are fertilized by peculiar antherozoids found near them.

There are two common genera,—Nitella and Chara. In Nitella the stems and branches are simple and naked; the leaves are in whorls of 5 to 8 and without stipules; the leaflets are large and often many-celled; the sporocarps arise singly or in clusters in the forkings of the leaves, and each has a crown of two superimposed whorls of five cells each. In Chara the stems and lower branches are usually corticated, i.e. there is a central tube surrounded by smaller tubes, sometimes spirally arranged, forming a cortex; the leaves are in whorls of 6 to 12, and usually with one or two stipules; the leaflets are always one-celled; the sporocarps arise from the upper side of the leaves, and each has a crown of one whorl of five cells. These plants exhibit beautifully the phenomenon of cyclosis, or circulation of protoplasm. Some species of Chara secrete calcium carbonate, and from this arises their popular name, "stone-worts."

CHAPTER XVIII.

FUNGI.

FUNGI are flowerless plants in which the special characteristic is the absence of chlorophyll and starch. Lacking these, they are unable to assimilate inorganic matter, and consequently live a saprophytic or a parasitic existence, that is, they live upon dead organic matter or in or upon some living host. They are essentially terrestrial plants, but some of them live a sort of semi-aquatic life.

Many very different forms are included among the Fungi. On the one hand there are microscopic forms,—and among them some authors include the bacteria, because they have no chlorophyll,—and on the other hand there are the mushrooms, etc., which are often of very large size. Fungi usually consist of two parts, the mycelium and the fruit. The mycelium is the vegetative portion of the plant. It is a mass of delicate, jointed, branched, colorless filaments intertwined to form a cottony or felty layer. It is the spawn of mushrooms and the common mold or mildew seen on decaying vegetable matter. The fruit consists of certain terminal mycelium filaments erected from the general mass and bearing spore-cells of various kinds. It is by differences in the method of fruiting or reproduction that the different fungi are distinguished from each other.

The Fungi, as a class, are of little importance in water

investigation. They are more often seen in sewage, and even there the number of important genera is small. For this reason a general classification of the Fungi is not given here, but simply a description of a few common genera.

ORDER SACCHAROMYCETES.

Saccharomyces.

Cells oval or somewhat rounded, colorless, with numerous vacuoles. They do not divide by the ordinary process of cell-division, but increase by a sort of sprouting or budding. A knob-like protuberance appears at one side of the cell; this increases in size and gradually assumes the form of the mother cell; it then separates and itself begins to bud, or it remains attached, forming a sort of irregular beading or branching. It does not develop true mycelia. It also reproduces by means of certain large cells whose protoplasm divides and forms several spores, sometimes called ascospores. There is no sexual reproduction. The Saccharomycetes are popularly called yeasts. They are well known for the alcoholic fermentation which they produce in sugar. The S. *cerevisiæ* is the common beer-yeast. Its cells average about 8 μ in diameter. There are other species which differ in the shape and size of the cells, in the character of the spores, in the temperature and time at which sprouting takes place, in the capacity to ferment sugars, in the time required to form yeast-films in the fermenting liquid, etc. (Pl. X, Fig. 4.)

ORDER ASCOMYCETES.

Penicillium.

This is the common "blue mold." The mycelium is composed of very many colorless, more or less branched filaments or hyphæ. The fertile hyphæ are erect and septate, and branch into a series of compound branches, each of which bears simple sterigmata upon which chains of oval conidia are borne. The most common species is *P. glaucum*. It has a pale bluish-green color. Its erect septate hyphæ are 1 to 2 mm. long, bearing a minute brush-like cluster of greenish conidia 2–4 μ in diameter. (Pl. X, Figs. 5 and 6.)

Aspergillus.
Mycelium as in Penicillium. Fertile hyphæ unseptate, swollen at apex (columella), bearing simple flask-shaped sterigmata, with chains of elliptical or spherical conidia. Often small yellowish or reddish bodies (perithecia or sclerotia) are found upon the sterile hyphæ at the base of the fertile branches. *A. repens* is a common species. The color is light greenish or brownish. Fertile hyphæ 2–4 mm. high, 10 μ diam.; columella 10–30 μ, head of conidia 100 μ, conidia 5 μ. (Pl. X, Fig. 7.)

ORDER PHYCOMYCETES.

FAMILY MUCORACEÆ.

Mucor.
Mycelium saprophytic or parasitic, richly branched, forming a felt-like layer. The hyphæ are seldom divided by septa. Conidia formed in sporangia which are spherical and borne on erect hyphæ. A common species is *M. racemosus*. Its sporangia are numerous, 20–70 μ in diameter, on the ends of long hyphæ. The spores are smooth, spherical, 4–8 μ in diameter. Secondary sporangia are sometimes seen on the main fruiting-branch. The color is whitish, and later a tawny brown. There are many other species, some of which produce alcoholic fermentation in sugar. (Pl. X, Fig. 8.)

FAMILY SAPROLEGNIACEÆ.

Saprolegnia.
Saprophytic or parasitic on plants or animals in water, sometimes producing pathogenic conditions, as, for example, in the "salmon-disease." They are often seen on dead flies, etc. The mycelium is composed of colorless or grayish hyphæ of large size attached to the substratum by root-like processes. The hyphæ are not constricted, as in Leptomitus. Sexual reproduction takes place by means of fertilized oospores. Asexual reproduction takes place by zoospores produced in special club-shaped zoosporanges which are borne terminally

upon certain hyphæ. The zoospores are numerous, sometimes in rows; they are bi-ciliated and motile even within the zoosporangium. After escaping from the zoosporangium they become covered with a thin membrane which they throw off before final swarming and germination. (Pl. XI, Fig. 1.)

Achlya.

Mycelium similar to that of Saprolegnia. The zoospores are non-motile when they escape from the zoosporangium. They arrange themselves in globular fashion outside the apex of the sporangium, assume a thin membrane, rest for a time, and ultimately escape, swim about, and germinate. (Pl. XI, Fig. 2.)

Leptomitus.

Hyphæ long, cylindrical, deeply constricted at intervals and at the base of the branches. Near the constriction there is usually a globular body, like an oil-globule. The grayish protoplasm is sometimes arranged in concentrated masses, and sometimes is uniformly distributed. The zoospores are formed in the interior of club-shaped terminal sporangia. They resemble those of Saprolegnia. Leptomitus is often found in masses in pipes conveying sewage or on the banks of polluted streams. (Pl. XI, Fig. 3.)

CHAPTER XIX.

PROTOZOA.

THE Protozoa are the lowest organisms belonging to the animal kingdom. The name Protozoa was used by the early writers to describe all minute organisms, whether animal or vegetable, but of late it has come to have a more definite meaning. It is now applied to those animal forms which are unicellular or multicellular by aggregation. Structurally the Protozoa are single cells, and where there is an aggregation of several cells each one preserves its identity. There is no differentiation, no difference in the function of the different cells. Thus, the Protozoa are definitely set off from the Metazoa or Enterozoa, which are multicellular, and which have two groups of cells, one group forming the lining to a digestive cavity and the other group forming the body-wall, which differ both in structure and in function. Most of the Protozoa are strictly unicellular.

It is extremely difficult to separate the unicellular Protozoa from the unicellular Protophyta. Theoretically there is a sharp distinction between the animal and vegetable kingdoms. Definitions may be found applicable to the higher types of life, but they overlap and become confused when applied to the lowest forms. For example, the fundamental difference betweeen the two kingdoms is supposed to lie in the phenomenon of nutrition. Plants can take up the carbon, oxygen, hydrogen, and nitrogen from mineral matter dissolved in water,—the nitrogen in the form of ammonia or nitrates, the carbon in the form of carbonic acid. Their food is in solution; hence they need no mouth or digestive apparatus. They absorb their nourishment through their entire surface. Animals, however, cannot take up nitrogen in a lower state than is found in the albumens, nor carbon except in combination with oxygen and hydrogen in the form of fat, sugar, starch, etc. The albumens and fats are not soluble in water; consequently the food of animals must consist of more or less solid particles. Animals therefore require a mouth, digestive cavity, organs for obtaining their food, etc. As albumens, fats, etc., are found in nature only as products of plant or animal life, it follows that all animal life is dependent upon vegetable or other animal life. There are, however, certain plants that live on organic matter (insectivorous plants, pitcher-plants) and even have digestive cavities, but all their relations show that they are real plants. There are other plants that are devoid of chlorophyll (Fungi), yet no one would think of calling them animals. Then there are many unicellular organisms that contain chlorophyll and have the vegetable, or holophytic, mode of nutrition, but that resemble the animal kingdom in other respects. Such, for example, are the Dinoflagellata and many of the green Flagellata. Because it is difficult to draw a sharp line between the vegetable and animal unicellular forms Haeckel proposed a new group, the Protista, lying between the two kingdoms. This group has been since known as the Phytozoa. The term is not used in this work, but the organisms have been placed in the one or the other of the two kingdoms according to the best available authority.

The Protozoan Cell.—The protozoan cell, or the individual protozoan, is a single mass of sarcode, or protoplasm, that possesses in a general way all the properties of the protoplasm of higher animal cells. It has a certain amount of irritability and movement, it assimilates food, it grows, and reproduces its kind. It is subject to the same chemical and physical reactions that are observed in higher forms. In size it varies from the tiniest corpuscle to a mass an inch in diameter. It is irregular in form, without a definite boundary; or it has a cell-wall and a definite symmetrical outline. Internally the cell usually contains a solid nucleus or a nuclear substance distributed through the cell and recognized by staining. It usually contains a contractile vacuole, which may be seen to expand and contract, discharging a watery or gaseous matter through the cell. There are also permanent vacuoles of watery fluid, gastric vacuoles formed by the water taken in with the food, oil-globules, and solid particles of starch, chlorophyll, etc. Externally there may be a cortical substance,—a denser layer of protoplasm giving definite shape to the cell—that is sometimes contractile. The exterior protoplasm may contain such secreted products as chitin, a nitrogenous horny matter, or cellulose, a non-nitrogenous substance, forming a cell-wall, cell-cuticle, or matrix. Substances may be deposited even outside of the protoplasmic layer. If perforated they are known as shells; if closed entirely, as cysts. Cysts are usually of a horny nature and are temporary products. External secretions of calcium carbonate, silicates, etc., are sometimes present.

The cell-protoplasm often exhibits certain internal flowing movements, described as the "streaming of the protoplasm." Portions of the protoplasm often extend outwards, forming processes. These are of two kinds, and the distinction between them has been used as a basis of classification. Those protozoa that have lobose, filamentous processes, known as pseudopodia, are called Myxopods; those that have motile hair-like processes, known as cilia or flagella, are called Mastigopods.

The simplest Protozoa absorb solid particles of food at any point on their surface. Digestion takes place within the cell. Protozoa higher in the scale of life have a distinct oral aperture through which the food enters, a sort of pharyngeal passage, and an anal aperture through which undigested portions of food are expelled. There is no real digestive cavity. Some Protozoa exhibit a simple kind of respiration. Experiment has shown that they take up oxygen and give out carbonic acid. Multiplication takes place by binary division, by encystment and spore-formation, by conjugation followed by spore-formation, or by conjugation followed by increased power of division. Strictly there is no sexual reproduction, though in certain instances there are processes corresponding to it.

Various classifications have been suggested for the Protozoa. None are entirely satisfactory. Bütschli has divided the Protozoa into four classes: the Sarcoda, Sporozoa, Mastigophora, and Infusoria. So far as fresh-water forms are concerned, the Sarcoda represent the Rhizopoda as described by Leidy. The Mastigophora and Infusoria are both included by the word Infusoria as used by Kent. Bütschli's classification with some modifications is given below, so far as it relates to the forms with which the water analyst is concerned. Many families and some entire orders are omitted.

CLASS RHIZOPODA.

Protozoa provided with variable, retractile root-like processes or pseudopodia; naked or enclosed in a carapace or external skeleton that is chitinous, calcareous, or siliceous; generally one and sometimes more than one nucleus; contractile vacuole present or absent.

There are five sub-classes—Lobosa, Reticularia, Heliozoa, Radiolaria, and Labyrinthulidea. The two latter are marine forms and therefore are omitted. The Lobosa and Reticularia are creeping animals; the Heliozoa are swimmers.

Sub-class Lobosa.

Rhizopoda in which the "amœba-phase" predominates in permanence and physiological importance. Pseudopodia lobose, not filamentous, arborescent, or reticulate. A denser external layer of protoplasm usually noticed. Provided with one or more nuclei and usually with a contractile vacuole. Reproduction commonly effected by simple fission, sometimes by a kind of budding.

Amœba.

A soft, colorless, granular mass of protoplasm; possessing extensile and contractile power; devoid of investing membrane, but having an external thickening of protoplasm; with variable, lobose, finger-like processes; ingesting food by flowing around and engulfing it; the absorbed food-material (diatoms, algæ, etc.) is often conspicuous. There are several species that vary in size and in the character of the pseudopodia. A common habitat is the superficial ooze of ponds or ditches. (Pl. XI, Fig. 4.)

Arcella.

An amœba-like organism enclosed in a chitinoid shell that is variable in shape, but more or less campanulate or dome-shaped, and that has a circular, somewhat concave base. When seen from above, it is disc-shaped, with a pale circular spot in the middle; when seen from the side, the upper surface is strongly convex. The shell usually has a brown color, and is sometimes smooth and sometimes hexagonally marked.

The protoplasmic mass occupies the central portion of the shell, but pseudopodia project through an opening in the concave base. There are many species, differing in shape and in the marks, ridges, etc., on the shell. *A. vulgaris* is the most common. (Pl. XI, Figs. 5 and 6.)

Difflugia.

Body enclosed in a spherical or pear-shaped membrane in which sand-grains, etc., are embedded. The lower part is sometimes prolonged as a neck, at the end of which is situated the mouth, through which finger-like pseudopodia may project. The surface of the shell is very rough and usually has a brownish or a gray color. Diatoms, etc., are frequently attached to the shell. The contained protoplasmic mass frequently has a green color, but the pseudopodia are colorless. There are several species, varying in shape and size. The diameter of Difflugia shells varies from 35 to 300 μ. (Pl. XI, Fig. 7.)

Sub-class Reticularia.

Rhizopoda covered with a secreted shell-like membrane with agglutinated particles of lime or sand. The projected pseudopodia are not finger-like, as in the Lobosa, but thread-like and delicately and acutely branched. The external denser layer of protoplasm is not as well marked as in the Lobosa. The shell is sometimes perforated by apertures.

Euglypha.

Body enclosed in a hyaline, ovoid shell, composed of regular hexagonal plates of chitinoid membrane, arranged in alternating longitudinal series. At the mouth the plates form a serrated margin. The upper portion of the shell is sometimes provided with spines. The protoplasm is almost entirely enclosed by the shell; the pseudopodia are delicate and branched. There are several species. (Pl. IX, Fig. 8.)

Trinema.

Body enclosed in a hyaline, pouch-like shell, with long axis inclined or oblique, and with mouth subterminal. Dome rounded; mouth inverted, circular, beaded at border. Pseudopodia as in Euglypha, but fewer in number. The two genera are quite similar, but Trinema is usually much smaller. One species. (Pl. XI, Fig. 9.)

SUB-CLASS HELIOZOA.

Rhizopoda generally spherical in form, with numerous radial, filamentous pseudopodia, which ordinarily exhibit little change of form, though they are elastic and contractile. Protoplasm richly vacuolated. One or more nuclei and contractile vacuoles. Chlorophyll grains sometimes present. Skeleton products sometimes present. The Heliozoa are generally found in fresh water. They are closely related to the marine Radiolaria.

Actinophrys.
A spherical mass of colorless protoplasm seemingly filled with small bubbles, with numerous long, fine rays springing from all parts of the surface. Contractile vesicle large and active. The organism moves with a slow gliding motion. It feeds on smaller protozoa, algæ-spores, etc. The most important species is *A. sol*, otherwise known as the "sun-animalcule." It is very common in swamp-water. (Pl. XI, Fig. 10.)

Heterophrys.
Like Actinophrys in general form, but with the body enveloped with a thick stratum of protoplasm defined by a granulated or thickly villous surface and penetrated by the pseudopodal rays.

CLASS MASTIGOPHORA.

Protozoa bearing one or more lash-like flagella, occasionally supplemented by cilia, pseudopodia, etc. With an indistinct, diffuse, or definite ingestive system, and usually with one or more contractile vesicles. Multiplication takes place by fission and by sporulation of the entire body-mass, the process often being preceded by conjugation of two or more zooids. The term Flagellata is used by some writers to describe this class of Protozoa.

Sub-class Flagellata.

Nucleated cells, with a definite, corticate, external layer of protoplasm and provided with one or more vibratile flagella. Food commonly ingested through an oral aperture in the cortical protoplasm, though some genera contain chlorophyll and are sustained by nutritional processes resembling those of plants. In some genera the cuticle is developed into stalks or collar-like outgrowths. Others produce chitinous shells or masses of jelly and are connected into arborescent or spherical colonies. Food-particles, starch-gains, chromatophore and chlorophyll corpuscles, oil-globules, pigment-spots (eye-spots) are often observed in the protoplasm of the cell.

The flagella of the Flagellata offer an interesting study. They are essentially different from cilia in their movement. Cilia are simply alternately bent and straightened. Flagella exhibit lashing movements to and fro and also throw themselves into serpentine waves. There are two kinds of flagella, distinguished by their movement—puisella and tractella. The former serve to drive the organism forward in the manner of a tadpole's tail. These are never found on the Flagellata. The tractellum is carried in front of the body and draws the organism after it, as a man uses his arms in swimming. The flagella of the Flagellata are always tractella.

Order Monadina.

Small, simple Flagellata, often naked or amœboid, usually colorless, seldom with chromatophores. With a single, large, anterior flagellum or sometimes with two additional flagella. Mouth area often wanting, never produced into a well-developed pharynx.

Family Cercomonadina.

Cercomonas.

Animalcules free-swimming, ovate or elongate, plastic, with a single long flagellum at anterior extremity and a caudal filament at the opposite extremity; no oral aperture. There are several species. Their length varies from 10 to 25 μ. (Pl. XII, Fig. 1.)

Family Heteromonadina.

Monas.

Very minute, free-swimming animalcules, colorless, globose or ovate, plastic, with no distinct cuticle; flagellum single, terminal; no distinct mouth. Several species, commonly found in vegetable infusions. Their length varies from 2 to 10 μ. They move with a "swarming" motion. (Pl. XII, Fig. 2.)

Anthophysa.

Animalcules colorless, obliquely pyriform, attached in spherical clusters to the extremities of slightly flexible, granular, opaque, more or less branching pedicles; two flagella, one longer than the other; no distinct mouth. In the common species, *A. vegetans*, the pedicle is dark brown and longitudinally striated. The detached stems somewhat resemble Crenothrix when observed with a low power. Zooids about 5 μ long; clusters 25 μ in diameter. Common in swamp-water. (Pl. XII, Fig. 3.)

Order Euglenoidea.

Somewhat large and highly developed monoflagellate forms, with firm, contractile, elastic cortical substance; some forms are stiff, others are capable of annular contraction and worm-like elongation. At the base of the flagellum there is a mouth leading into a pharyngeal tube, near which is a contractile vacuole. Rarely with two flagella.

FAMILY CŒLOMONADINA.

Cœlomonas.

Animalcules free-swimming, monoflagellate, highly contractile and variable in form, with distinct oral aperture and a spheroidal pharyngeal chamber; nucleus and contractile vacuole conspicuous; no trichocysts; with innumerable green chlorophyll granules. Nutrition largely vegetal. One species. Length about 50 μ. (Pl. XII, Fig. 4.)

Raphidomonas (Gonyostomum).

Animalcules free-swimming; ovate-elongate, flexible body, widest anteriorly and tapering posteriorly, two to three times as long as wide; two flagella, one of them trailing; oral aperture at anterior end conducts to a conspicuous triangular or lunate pharyngeal chamber; contractile vacuole conspicuous; nucleus ovate; a brownish germ-sphere posteriorly located; many large bright green chlorophyll bodies; numerous rod-like bodies called trichocysts; oil-globules often present. Length 40 to 70 μ. Reproduction by spores formed in the germ-sphere. One species, *R. semen*. The genus Trentonia, described by Dr. A. C. Stokes, is similar to Raphidomonas except that it has no trichocysts. (Pl. XII, Fig. 5.)

FAMILY EUGLENINA.

Euglena.

Free-swimming animalcules, fusiform or elongate, exceedingly flexible in form; with highly elastic cuticle terminating posteriorly in a tail-like prolongation; endoplasm bright green or reddish; flagellum flexible, issuing from an anterior notch at the bottom of which is the oral aperture and a red pigment-spot. There are several common species. *E. viridis* is the most common. It is often found in immense numbers in stagnant pools, forming a characteristic green or reddish scum. Length varies from 40 to 150 μ. *E. acus* is an elongated form with tapering ends. It is longer than *E. viridis*, but less broad. It is also less variable in form. *E. deses* is a very long cylindrical form. (Pl. XII, Fig. 6.)

Trachelomonas.
Monoflagellate animalcules, changeable in form, enclosed within a free-floating, spheroidal, indurated sheath or lorica; flagellum protruded through an aperture in the lorica. The color of the animalcule is green, with a red pigment-spot; the color of the lorica is generally a reddish brown. There are several species. Diameter of lorica generally about 25 μ. (Pl. XII, Fig. 7.)

Phacus.
Free-swimming animalcules; form persistent, leaf-like, with sharp-pointed, tail-like prolongation; terminal oral aperture and tubular pharynx; flagellum long, vibratile; surface indurated; endoplasm green, with red pigment-spot; contractile vacuole large, subspherical. Length about 50 μ, but quite variable. (Pl. XII, Fig. 8.)

Order Isomastigoda.

Small and middle-sized forms of monaxonic, rarely bilateral shape. Fore end with two or more flagella. Some are colored, some colorless; naked or with strong cuticle or secreting an envelope. Nutrition generally holophytic (i.e. like a green plant).

Family Chrysomonadina.

Synura.
Free-swimming animalcules, united in subspherical social clusters, each zooid contained in a separate membranous sheath or lorica, the posterior extremities of which are stalk-like and confluent; two subequal flagella, sometimes long; pigment-spots minute or absent; two brown color-bands produced equally throughout the length of the two lateral borders; a vacuolar space at the anterior extremity and several contractile vacuoles; oil-globules often observed. Length of individual zooids about 35 μ; diameter of clusters varies from 30 to 100 μ. There is

one species, *S. uvella*, with several varieties. The colonies move with a brisk rolling motion, caused by the combined action of the flagella. Common in swamp-waters. (Pl. XII, Fig. 9.)

Uvella.

An uncertain genus. Uvella differs from Synura in the non-possession of a separate investing membrane or lorica and by the posterior location of the contractile vacuole. There are usually few zooids in the cluster. (Pl. XII, Fig. 10.)

Syncrypta.

Free-swimming animalcules, united into spherical clusters as in Synura, without lorica, but with the entire colony immersed within a gelatinous matrix, beyond the periphery of which the flagella alone project; two subequal flagella; brownish lateral color-bands evenly developed; one or two pigment-spots; contractile vacuole between the color-bands. Length of zooids about 10 μ. Diameter of colony about 50 μ, including gelatinous zoogloea. There is but one species, *S. volvox*. It resembles Synura. It is not common. Pl. XII, Fig. 11.)

Uroglena.

Animalcules forming almost colorless spheroidal colonies barely visible to the naked eye. The matrix of the colony is a transparent gelatinous shell filled with a watery substance. The zooids are embedded on the periphery, with their flagella extending outwards and by their vibration causing the colony to revolve. The zooids are pyriform, with anterior border rounded and truncated, tapering posteriorly and sometimes continued backwards as a contractile thread; with two light yellowish-green pigment-bands; one eye-spot at the base of the flagella; two unequal flagella; one or more contractile vacuoles; oil-globules and a large amylaceous body often present. Length of zooids is about 6 to 12 μ. The colonies are from 200 to 500 μ in diameter. There are several rather indistinct species. The zooids multiply by division into twos or fours. The colonies also divide, a hollow first appearing on one side, followed by a rounding at the two poles and a subsequent twisting apart. The Uroglena colonies are very fragile. (Pl. XII, Figs. 12 and 13.)

Dinobryon.
Animalcules with urn- or trumpet-shaped loricæ attenuated posteriorly and set one into another so as to form a compound branching polythecium. The zooids are elongate-ovate, attached to the bottom of the loricæ by transparent elastic threads; two unequal flagella; two brownish or greenish lateral color-bands; a conspicuous pigment-spot; nucleus and contractile vacuole sub-central. The polythecium is constructed through the successive terminal gemmation of the zooids. Length of separate loricæ 15 to 60 μ. The polythecium may contain from 2 to 500 loricæ. The usual number is between 25 and 50. Reproduction takes place by spore-formation. The spores sometimes remain attached to the polythecium, or they may become scattered. When free they are liable to be mistaken for small Cyclotella. The spores are from 8 to 20 μ in diameter. There are several species. *D. sertularia* is the most common. (Pl. XIII, Fig. 1.)

Cryptomonas.
Free-swimming animalcules, illoricate, but persistent in form, ovate or elongate, compressed asymmetrically; flagella two, long, equal in length, issuing from a deep groove or furrow; large oral aperture at the base of the flagella continued backwards as a tubular pharynx; two lateral bright green color-bands; conspicuous nucleus and contractile vacuole; oil-globules often present. Length from 40 to 60 μ. (Pl. XIII, Fig. 2.)

Mallomonas.
Free-swimming animalcules, oval or elliptical, persistent in shape; surface covered with overlapping horny plates from which arise long hair-like setæ; under low power the surface has a crenulated appearance. One long, slender anterior flagellum; indistinct contractile vacuole. Endoplasm vacuolar, greenish or yellowish. Length from 20 to 40 μ. (Pl. XIII, Fig. 3.)

FAMILY CHLAMYDOMONADINA.—This family is often referred to the vegetable kingdom.

Chlamydomonas.
Animalcules ovate, with two or more flagella, one large green

color-mass, a delicate membranous shell, usually with a pigment-spot and one or more contractile vacuoles. The protoplasm divides into new individuals within the envelope. Length from 10 to 30 μ. (Pl. XIII, Fig. 4.)

FAMILY VOLVOCINA.—Often included under Protozoa. See page 193.

SUB-CLASS CHOANOFLAGELLATA.

Mastigophora provided with an upstanding collar surrounding the anterior pole of the cell, from which the single flagellum springs. (Omitted from this work.)

SUB-CLASS DINOFLAGELLATA.

Mastigophora are characterized by the presence of a longitudinal groove, marking the anterior region and the ventral surface, and from which a long flagellum projects. In every genus but one there is also a transverse groove in which lies horizontally a second flagellum, at one time mistaken for a girdle of cilia. The animalcules are bilaterally asymmetrical. They are occasionally naked, but most genera are covered with a cuticular shell of cellulose, either entire or built of plates. The endoplasm contains chlorophyll, starch-granules, and a brown coloring matter similar to that of diatoms. The nucleus is large and branching. There is no contractile vacuole. Multiplication takes place by transverse binary fission.

Because of the presence of the cellulose shell, chlorophyll, starch-granules, and a holophytic (vegetal) mode of nutrition the Dinoflagellata are often classed in the vegetable kingdom. Many of the Dinoflagellata are marine forms. Some are phosphorescent.

Peridinium.
Free-swimming animalcules enclosed within a cellulose shell composed of polygonal facets. With a high power the facets exhibit a delicate reticulation. A transverse groove divides the body into two subequal parts. A second groove extends from the first towards the apical extremity. Two flagella, one in the transverse groove, the other proceeding from the junction of the two grooves. Color yellowish green or brown. There are one or more pigment-spots. Length from 40 to 75 μ. There are several species. *P. tabulatum* is the most common. (Pl. XIII, Fig. 5.)

Ceratium.
Free-swimming animalcules enclosed within a shell consisting of two subequal segments, one or both of which are produced into conspicuous horn-like prolongations, often covered with tooth-like processes. There is a central transverse furrow and a second groove extending from the centre of the ventral aspect towards the anterior pole. Two flagella, one of which lies in the transverse groove. The brown color is not as marked as in Peridinium. Length from 25 to 150 μ. There are several species, varying considerably in the character of the horn-like projections. (Pl. XIII, Fig. 6.)

Glenodinium.
Free-swimming animalcules covered with a smooth, cellulose shell not made up of facets, consisting of two subequal parts. There is a conspicuous transverse groove and a much less conspicuous secondary groove. Two typical flagella. Body ovate. Color brownish. Pigment-spot sometimes present. Length about 40 to 55 μ. Glenodinium is often surrounded by a wide, irregular mass of jelly. (Pl. XIII, Fig. 7.)

Gymnodinium.
Quite similar to Peridinium, but without a protecting shell.

SUB-CLASS CYSTOFLAGELLATA.

Marine forms.

CLASS INFUSORIA.

In its broadest sense the word Infusoria includes all the Protozoa except the Rhizopoda and Sporozoa. As used here, following Bütschli, it includes only the Ciliata and Suctoria.

Sub-class Ciliata.

Protozoa of relatively large size, furnished with cilia, but not with flagella. The cilia occur as a single band surrounding the oral aperture or are dispersed over the entire body. Modification of the cilia into setæ or styles is sometimes observed. There is generally a well-developed oral and anal aperture. The nucleus varies in different genera. Besides one larger, oblong nucleus a smaller one (paranucleus) is often present. One or more contractile vacuoles present. They all possess a delicate but well-defined ectoderm, elastic, but constant in form. They occur naked or enclosed in horny or siliceous shells or in gelatinous envelopes. Some genera are stalked. Multiplication takes place by transverse fission. Conjugation has been observed, but the part that it plays in the life-history is not well known. Many of the Ciliata are parasites in higher animals.

The Ciliata are divided into four orders according to the character and distribution of their cilia.

Order Hypotricha.

Ciliata in which the body is flattened and the locomotive cilia are confined to the ventral surface, and are often modified and enlarged to the condition of muscular appendages. Usually an adoral band of cilia, like that of Heterotricha. Dorsal surface smooth or provided with tactile hairs only. Mouth and anus conspicuous.

Euplotes.
Animalcules free-swimming, encuirassed, elliptical or orbicular, with sharp laminate marginal edges, and usually a plane ventral, and convex, sometimes furrowed, dorsal surface. Peristome-field arcuate, extending backwards from the frontal border to or beyond the centre of the ventral surface, sometimes with a reflected and ciliate inner border. Frontal styles six or seven in number; three or more irregularly scattered ventral styles, and five anal styles; four isolated caudal styles along the posterior margin. Endoplast linear. Single spherical contractile vesicle near anal aperture. Length about 125 μ. (Pl. XIII, Fig. 8.)

Order Peritricha.

Ciliata with the cilia arranged in one anterior circlet or in two, an anterior and a posterior; the general surface of the body destitute of cilia. The Peritricha are sometimes divided into two suborders, the free-swimming forms and the attached forms.

Halteria.
Animalcules free-swimming, colorless, more or less globose, terminating posteriorly in a rounded point. Oral aperture terminal, eccentric, associated with a spiral or subcircular wreath of large cirrose cilia. A zone of long hair-like setæ or springing-hairs developed around the equatorial region, the sudden flexure of which appendages enables the organism to progress through the water by a series of leaping movements, in addition to their ordinary swimming motions. Length 15 to 30 μ. There are several species, some of them colored green. (Pl. XIII, Fig. 9.)

Vorticella.
Animalcules ovate, spheroidal, or campanulate, attached posteriorly by a simple undivided, elongate and contractile, thread-like pedicle; the pedicle enclosing an elastic, spirally disposed, muscular fibrilla, and assuming suddenly on contraction a much-shortened and usually corkscrew-like contour.

Adoral system consisting of a spirally convolute ciliary wreath, the right limb of which descends into the oral cleft, the left one obliquely elevated and encircling the ciliary disk. The entire adoral wreath contained within and bounded by a more or less distinctly raised border—the peristome—between which and the elevated ciliary disk, on the ventral side, the widely excavated cleft or vestibulum is situated. The vestibulum is continued further into a conspicuous cleft-like pharynx, and terminates in a narrow tubular œsophagus. Anal aperture opening into the vestibulum. Contractile vesicle single, spherical, near the vestibulum. Nucleus elongate. Multiplication by longitudinal fission, by gemmation, and by the development of germs. There exists a very large number of species, varying considerably in size and shape. The length varies from 25 to 200 μ. Vorticella are often found floating in water attached to masses of Anabæna, etc. (Pl. XIII, Fig. 10.)

Zoothamnium.

Animalcules structurally identical with those of Vorticella, ovate, pyriform, or globular, often dissimilar in shape and of two sizes, stationed at the extremities of a branching, highly contractile pedicle or zoodendrium. Numerous species.

Epistylis.

Animalcules campanulate, ovate, or pyriform, structurally similar to Vorticella, attached in numbers to a rigid, uncontractile, branching, tree-like pedicle or zoodendrium; the zooids usually of similar size and shape. Numerous species. (Pl. XIII, Fig. 11.)

Order Heterotricha.

Ciliata possessing two distinct systems of cilia, one a band or spiral or circlet of long cilia developed in the oral region, the other composed of short, fine cilia covering the entire body. The cortical layer is usually highly differentiated.

Tintinnus.

Animalcules ovate or pyriform, attached posteriorly by a

slender retractile pedicle within an indurated sheath or lorica. The shape of the lorica is generally cylindrical; it is free-floating; it is somewhat mucilaginous and attracts to its outer surface foreign particles, such as grains of inorganic matter, diatom-shells, etc. The peristome-field of the organism occupies the entire anterior border, circumscribed by a more or less-complex circular or spiral wreath of long, powerful, cirrose cilia, the left limb or extremity of which is spirally involute and forms the entrance to the oral fossa. This fossa is continued as a short, tubular pharynx. Anus posteriorly situated, subterminal. Cuticular cilia very fine, distributed evenly throughout, clothing both the body and the retractile pedicle. Length of lorica 80 to 150 μ. There are many species, varying greatly in the size and shape of the loricæ. In the fresh-water forms the lorica is generally cylindrical. Another genus, Tintinnidium, varies from Tintinnus only in having a more mucilaginous sheath and in being permanently attached to foreign objects. (Pl. XIII, Fig. 12.)

Codonella.
Animalcules conical or trumpet-shaped, solitary, free-swimming, highly contractile, inhabiting a helmet- or bell-shaped lorica, to which they are attached by their posterior extremity. The anterior region truncate or excavate, forming a circular peristome having an outer fringe of about twenty long, tentacle-like cilia, and an inner collar-like border, or frill, which bears an equal number of slender, lappet-like appendages. Entire cuticular surface clothed with fine, vibratile cilia. Lorica not perforated, of chitinous consistence, often of a brown color, sometimes sculptured or mixed with foreign granular substances. Length of lorica 50 to 150 μ. Several species, mostly marine. (Pl. XIV, Fig. 1.)

Stentor.
Animalcules sedentary or free-swimming at will; bodies highly elastic and variable in form: when swimming and contracted, clavate, pyriform, or turbinate; when fixed and extended, trumpet-shaped, broadly expanded anteriorly, tapering off and attenuated towards the attached posterior extremity. Peristome

describing an almost complete circuit around the expanded anterior border, its left-hand extremity or limb spirally involute, forming a small pocket-shaped fossa conducting to the oral aperture, the right-hand limb free and usually raised considerably above the opposite or left-hand one. Peristomal cilia cirrose, very large and strong; cilia of the cuticular surface very fine, distributed in even longitudinal rows, occasionally supplemented by scattered hair-like setæ. Nucleus band-like, moniliform, or rounded. Contractile vesicle complex. Multiplication by oblique fission and by germs separated from the band-like endoplast. There are many species, some of large size, colorless, or greenish, bluish, brownish, etc. (Pl. XIV, Fig. 2.)

Bursaria.

Animalcules free-swimming, broadly ovate, somewhat flattened on one side, anteriorly truncate. Peristome-field pocket-shaped, deeply excavate, situated obliquely on the anterior half of the body, having a broad oral fossa in front, and a cleft-like lateral fissure, which extends from the left corner of the contour border to the middle of the ventral side ; no tremulous flap. Pharynx long, funicular, bent towards the left, and forming a continuation of the peristome excavation. Adoral ciliary wreath broad, much concealed, lying completely within the peristome-cleft. Cuticular cilia fine, in longitudinal rows. Anus posteriorly situated, terminal. Nucleus band-like, curved, or sinuous. Contractile vesicles distinct, usually multiple. Few species. Length 300 to 500 μ. (Pl. XIV, Fig. 3.)

Order Holotricha.

Ciliata with but one sort of cilia, these covering the body uniformly and almost completely. A variously modified extensile or undulating membrane sometimes present. Oral and anal orifices usually conspicuous. Trichocysts sometimes present in the cuticular layer.

Paramæcium.

Animalcules free-swimming, ovate or elongate, asymmetrical,

more or less flexible, but persistent in shape. Finely ciliated throughout, the cilia of the oral region not differing in size or character from those of the general surface of the body. An oblique groove developed on the ventral surface, at the posterior extremity of which is situated the oral aperture. Cortical layer usually enclosing trichocysts. Contractile vesicles and nucleus conspicuous, the former sometimes stellate. There are several species. The most important is *P. aurelia*, which is often found in sewage-polluted and stagnant water. It is colorless, has a length of about 225 μ, and moves with a brisk rotatory motion. (Pl. XIV, Fig. 4.)

Nassula.
Animalcules ovate, cylindrical, flexible, but not polymorphic, usually highly colored—rose, red, blue, yellow, etc. Oral aperture lateral. Pharynx armed with a simple horny tube or with a cylindrical fascicle of rod-like teeth. Entire surface of cuticle finely and evenly ciliate. The cortical layer sometimes containing trichocysts. There are several species, varying in color, shape, and size. Length 50 to 250 μ. (Pl. XIV, Fig. 5.)

Coleps.
Animalcules ovate, cylindrical, or barrel-shaped, persistent in shape, cuticular surface divided longitudinally and transversely by furrows into quadrangular facets; these facets are smooth and indurated, the narrow furrows soft and clothed with cilia; the anterior margin mucronate or denticulate; the posterior extremity mucronate and provided with spines or cusps. Oral aperture apical, terminal, surrounded with cilia. Anal aperture at posterior extremity. Color gray or light brown. The most common species is *C. hirtus*, which has a length of about 60 μ. (Pl. XIV, Fig. 6.)

Enchelys.
Animalcules free-swimming, elastic, and changeable in shape, pyriform or globose. Oral aperture situated at the termination of the narrower and usually oblique truncate anterior extremity. Anal aperture at the posterior termination. Cuticular surface finely and entirely ciliate; the cilia are longer in the region

of the mouth. Few species. Length about 25 to 50 μ.
(Pl. XIV, Fig. 7.).

Trachelocerca.

Animalcules colorless, highly elastic, and changeable in form, the anterior portion produced as a long, flexible, narrow, neck-like process, the apical termination of which is separated by an annular constriction from the preceding part, and is perforated apically by the oral aperture. Cuticular surface evenly and finely ciliate; a circle of larger cilia developed around the oral region. Length of extended body about 150 μ. Few species. (Pl. XIV, Fig. 8.)

Pleuronema.

Animalcules ovate, colorless. Oral aperture situated in a depressed area near the centre of the ventral surface, supplemented by an extensile, hood-shaped, transparent membrane or velum, which is let down or retracted at will. Numerous longer vibratile cilia stationed at the entrance of the oral cavity. The general surface of the body clothed with long, stiff, hair-like setæ. The cortical layer usually containing trichocysts. Length 60 to 100 μ. Few species. (Pl. XIV, Fig. 9.)

Colpidium.

Animalcules free-swimming, colorless, kidney-shaped. Entirely ciliate. Oral aperture inferior, subterminal. Pharynx supported throughout its length by an undulating membrane which projects exteriorly in a tongue-like manner. Two nuclei, rounded, sub-central. Length 50 to 100 μ. One species. (Pl. XV, Fig. 1.)

Sub-class Suctoria (Tentaculifera or Acinetaria).

Protozoa with neither flagellate appendages nor cilia in their adult state, but seizing their food and effecting locomotion, when unattached, by means of tentacles. These are simply adhesive or tubular and provided at their distal extremity with a cup-like sucking-disk. Nucleus usually much branched. One or more contractile vesicles. Multiplication by longitudinal or transverse fission or by external or internal bud-formation. The young forms are ciliated. Most of the Suctoria are sedentary.

Acineta.
Animalcules solitary, ovate or elongate, secreting a protective lorica, to the sides of which they are adherent or within which they may remain freely suspended. Lorica transparent, triangular or urn-shaped, supported upon a rigid pedicle. Tentacles suctorial, capitate, distributed irregularly or in groups. There are many species. (Pl. XV, Fig. 2.)

CHAPTER XX.

ROTIFERA.

The Rotifera, or Rotatoria, comprise a well-defined group of minute multicellular animals. They are often included among the Vermes, but some of them possess characteristics that suggest the Arthropoda.

Though microscopic in size, the Rotifera are quite highly organized. They have a well-defined digestive system, including a mouth, or buccal orifice; a mastax, a peculiar set of jaws for mastication; salivary glands; an œsophagus; gastric glands; a stomach; an intestine; and an anus. There is a vascular system, a muscular system, and, it is claimed, a nervous system. There is a conspicuous reproductive system, and both males and females are observed, although the males are rare. The transparency of most of the Rotifera renders these various organs subjects of easy investigation.

The organisms are protected by a firm, homogeneous, structureless cuticle, often hardened by a development of chitin, forming a carapace or lorica. Some genera are further protected by an exterior casing or sheath, called an "urceolus," which may be gelatinous and transparent, as in Floscularia, or covered with foreign particles or pellets, as in Melicerta.

The Rotifera are generally bilaterally symmetrical, with a dorsal and ventral surface, with definite head region and tail region, broadest anteriorly and tapering posteriorly. There are three features of the Rotifera that deserve special attention, partly because they are unique in the organisms of this group and partly because they are used as the basis of classification. They are the ciliary wreath, the mastax, and the foot.

The ciliary wreath consists of one or more circlets of cilia springing from disc-like lobes surrounding the mouth at the anterior end. By their continual lashing they present the appearance of wheels, giving to these organisms the name of "wheel-animalcules." Their function is to assist in locomotion, to create currents in the water by which food-particles are carried into the mouth, and to conduct this food-material through the alimentary canal. The disc-like lobe bearing the cilia is known by the names of corona, trochal disc, or velum. It takes different shapes in different rotifers. Its simplest form is an oval or circle. In more complex forms it is intricately folded, as shown on Pl. XVI, Figs. A to E. The ciliated wreath is often supplemented by certain projecting processes, ciliated or bearing setæ or bristles.

The foot, pseudopodium, or posterior extremity of a rotifer presents several different types. It may be fleshy and transversely wrinkled, or hard and jointed; it may be non-retractile or retractile; often the jointed forms are telescopic; it may terminate in a sort of sucking-disc or in a ciliated expansion, or it may be furcate, or divided into toes, as shown on Pl. XVI, Figs. F to I. In some species the foot is altogether lacking.

The mastax is a sort of muscular bulb forming a part of the pharynx and containing the trophi. It has an opening above from the mouth and below into the œsophagus. The trophi, or teeth, are peculiar calcareous structures. Their function is to grind the food before it passes into the stomach, and this grinding movement may be witnessed through the transparent walls of many rotifers. The trophi consist of two toothed, hammer-like bodies, or mallei, that pound on a sort of split anvil, or incus. The malleus consists of an upper part, the head or uncus, and a lower part, the handle or manubrium. The incus also consists of two parts, a symmetrically divided upper part, the rami, that receives the blow of the malleus, and a lower part or fulcrum. The trophi show great modifications in different genera in the shape and proportion of the various parts.* Pl. XVI, Fig. J represents a typical form.

These three characteristics—the arrangement of the ciliary wreath, the structure of the foot, and the form of the trophi —serve as the basis for dividing the Rotifera into orders and families. The following classification is that adopted by Hudson and Gosse. Only the typical and very common genera are described.

---

*The following terms are used to describe the trophi (see Pl. XVI, Figs. J to P):

Malleate.—Mallei stout; manubria and unci of nearly equal length; unci 5- to 7-toothed; fulcrum short.

Submalleate.—Mallei slender; manubria about twice as large as the unci; unci 3- to 5-toothed.

Forcipitate.—Mallei rod-like; manubria and fulcrum long; unci pointed or evanescent; rami much developed and used as forceps.

Incudate.—Mallei evanescent; rami highly developed into a curved forceps; fulcrum stout.

Uncinate.—Unci 2-toothed; manubria evanescent; incus slender.

Ramate.—Rami subquadrate, each crossed by two or three teeth; manubria evanescent: fulcrum rudimentary.

Malleo-ramate.—Mallei fastened by unci to rami; manubria three loops soldered to the unci; unci 3-toothed; rami large, with many striæ parallel to the teeth; fulcrum slender.

ORDER RHIZOTA.

Rotifera fixed when adult; usually inhabiting a gelatinous tube excreted from the skin. Foot transversely wrinkled, not contractile within the body, ending in an adhesive sucking-disc or cup, without telescopic joints, never furcate.

FAMILY FLOSCULARIADÆ.—Corona produced longitudinally into lobes bearing the setæ. Mouth central. Ciliary wreath a single half-circle above the mouth. Trophi uncinate.

Floscularia.
Frontal lobes short, expanded, or wholly wanting. Setæ very long and radiating, or short and cilia-like. Foot terminated by a non-retractile peduncle, ending in an adhesive disc. Inhabiting a transparent gelatinous tube into which the animal contracts when alarmed. There are several species, varying in length from 200 to 2500 μ. (Pl. XV, Fig. 3.)

FAMILY MELICERTADÆ.—Corona not produced in lobes bearing setæ. Mouth lateral. Ciliary wreath a marginal continuous curve bent on itself at the dorsal surface so as to encircle the corona twice, with the mouth between its upper and lower curves, and having a dorsal gap between its points of flexure. Trophi malleo-ramate.

Melicerta.
Corona of four lobes. Dorsal gap wide. Dorsal antennæ minute. Ventral antennæ obvious. Inhabiting tubes built up of pellets. Length 800 to 1500 μ. Few species. *M. ringens* is very common on water-plants. (Pl. XV, Fig. 4.)

Conochilus.
Corona horseshoe-shaped, transverse; gap in ciliary wreath ventral. Mouth on the corona, and towards its dorsal side. Dorsal antennæ very minute or absent. Ventral antennæ obvious. Forming free-swimming clusters of several individuals, inhabiting coherent gelatinous tubes. Length 500 to 1200 μ. Two species. *C. volvox* is very common. (Pl. XV, Fig. 5.)

ORDER BDELLOIDA.

Rotifera that swim with their ciliary wreath and creep like a leech. Foot wholly retractile within the body, telescopic, at the end almost invariably divided into three toes.

FAMILY PHILODINADÆ.—Corona a pair of circular lobes transversely placed. Ciliary wreath a marginal continuous curve bent on itself at the dorsal surface so as to encircle the corona twice, with mouth between its upper and lower curves, and having also two gaps, the one dorsal between its points of flexure, the other ventral in the upper curve opposite to the mouth. Trophi ramate.

Rotifer.

Eyes two, within the frontal column. The most common species is *R. vulgaris*, which has a white body, smooth, and tapering to the foot. Spurs and dorsal antennæ of moderate length. Length about 500 μ. This was one of the first rotifers discovered. It gave its name to the entire class. (Pl. XV, Fig. 6.)

ORDER PLOIMA.

Rotifera that swim with their feet and (in some cases) creep with their toes. This is the largest and most important order of Rotifera.

SUB-ORDER ILLORICATA.

Integument flexible, not stiffened to an enclosing shell. Foot, when present, almost invariably furcate, but not transversely wrinkled; rarely more than feebly telescopic, and partially retractile.

FAMILY MICROCODIDÆ.—Corona obliquely transverse, flat, circular. Mouth central. Ciliary wreath a marginal continuous curve encircling the corona, and two curves of larger cilia, one on each side of the mouth. Trophi forcipitate. Foot stylate.

Microcodon.

Eye single, centrally placed, just below the corona. One

species. Length about 200 μ, of which the foot is more than half. (Pl. XV, Fig. 7.)

FAMILY ASPLANCHNADÆ.—Corona subconical, with one or two apices. Ciliary wreath single, edging the corona. Intestine and cloaca absent.

Asplanchna.
Corona with two apices. Trophi incudate, not enclosed within a mastax. Stomach of moderate size, spheroidal. Viviparous. Several species. Very large and transparent. (Pl. XV, Fig. 8.)

FAMILY SYNCHÆTADÆ.—Corona a transverse spheroidal segment, sometimes much flattened, with styligerous prominences. Ciliary wreath a single interrupted or continuous marginal curve encircling the corona. Mastax very large, pear-shaped. Trophi forcipitate. Foot minute, furcate.

Synchæta.
Form usually that of a long cone whose apex is the foot ; front furnished with two ciliated club-shaped prominences. Ciliary wreath of interrupted curves. Foot minute, furcate. Several species. Length 150 to 300 μ. (Pl. XVI, Fig. 1.)

FAMILY TRIARTHRADÆ.—Body furnished with skipping appendages. Corona transverse. Ciliary wreath single, marginal. Foot absent.

Polyarthra.
Eye single, occipital. Mastax very large and pear-shaped. Trophi forcipitate. Provided with two clusters of six spines on the shoulders, the spines being in the form of serrated blades. Length about 125 μ. (Pl. XVI, Fig. 2.)

Triarthra.
Eyes two, frontal. Mastax of moderate size. Trophi malleoramate. Spines single, two lateral, one ventral. There are three species, differing chiefly in the length of the spines. In the most common species the spines are twice the length of the body. Length of body about 150 μ. (Pl. XVI, Fig. 3.)

FAMILY HYDATINADÆ.—Corona truncate, with styligerous prominences. Ciliary wreath two parallel curves, the one marginal fringing the corona and mouth, the other lying within the first, the styligerous prominences lying between the two. Trophi malleate. Foot furcate.

Hydatina.
>Body conical, tapering towards the foot. Foot short and confluent with the trunk. Eye absent. This is one of the largest of the Ploima. Length about 600 μ.

FAMILY NOTOMMATADÆ.—Corona obliquely transverse. Ciliary wreath of interrupted curves and clusters, usually with a marginal wreath surrounding the mouth. Trophi forcipitate. Foot furcate. This family is the most typical, the most highly organized, of the Rotifera.

Diglena.
>Body subcylindrical, but very versatile in outline, often swelling behind and tapering to the head. Eyes two, minute, situated near the edge of the front. Foot furcate. Trophi forcipitate, generally protrusile. Several species. Length 125 to 400 μ. (Pl. XVI, Fig. 4.)

SUB-ORDER LORICATA.

Integument stiffened to a wholly or partially enclosing shell; foot various.

FAMILY RATTULIDÆ.—Body cylindrical or fusiform, smooth, without plicæ or angles; contained in a lorica closed all around, but open at each end, often ridged. Trophi long, asymmetrical. Eye single, cervical.

Mastigocerca.
>Body fusiform or irregularly thick, not lunate. Toe a single style, with accessory stylets at its base. Lorica often furnished with a thin dorsal ridge. Many species. (Pl. XVI, Fig. 5.)

FAMILY COLURIDÆ.—Body enclosed in a lorica, usually of firm consistence, variously compressed or depressed, open at both ends, closed dorsally, usually open or wanting ventrally. Head surrounded by a chitinous arched plate or hood. Toes two, rarely one, always exposed.

Colurus.
Body subglobose, more or less compressed. Lorica of two lateral plates, open in front, gaping behind. Frontal hood in form of a non-retractile hook. Foot prominently extruded, of distinct joints, terminated by two furcate toes. Many species.

FAMILY BRACHIONIDÆ.—Lorica box-like, open at each end, generally armed with anterior and posterior spines. Foot very long, flexible, uniformly wrinkled, without articulation; toes very small.

Brachionus.
Lorica without elevated ridges, gibbous both dorsally and ventrally. Foot very flexible, uniformly wrinkled, without articulation; toes very small. Free-swimming. Many species. (Pl. XVII, Fig. 1.)

Noteus.
Lorica facetted and covered with raised points; gibbous dorsally, flat ventrally. Foot obscurely jointed. Toes moderately long. Eyes wanting. Length 350 μ.

FAMILY ANURÆADÆ.—Lorica box-like, broadly open in front, open behind only by a narrow slit. Usually armed with spines or elastic setæ. Foot wholly wanting.

Anuræa
Lorica an oblong box, open widely in front, narrowly in rear; dorsal surface usually tessellated. The occipital ridge always, the anal sometimes, furnished with spines. The egg after extrusion is carried attached to the lorica. Free-swimming. Length about 125 μ. (Pl. XVII, Figs. 2 and 3.)

Notholca.
Lorica ovate, truncate and six-spined in front, sometimes produced behind; of two spoon-like plates united laterally. No posterior spines. Dorsal surface marked longitudinally with alternate ridges and furrows. Expelled egg not usually carried. Free-swimming. Several species. (Pl. XVII, Fig. 4.)

ORDER SCIRTOPODA.

Rotifera swimming with their ciliary wreath and skipping with arthropodous limbs; foot absent. There is but one genus, Pedalion, and that is rare.

CHAPTER XXI.

CRUSTACEA.

THE Crustacea belong to the Arthropoda—that is, to that group of the Articulates that have jointed appendages. Most of the larger Crustacea are marine, but many of the smaller forms are found in fresh water. These vary in size from objects barely visible to the naked eye to bodies several centimeters in length. The most common forms are somewhat less in size than the head of a pin.

The fresh-water Crustacea have been sometimes divided into two groups, the Entomostraca and the Malacostraca.

The Malacostraca are comparatively large forms. They include the Amphipoda, one of which is Gammarus pulex, the "water-crab"; the Isopoda, with Asellus aquaticus, or the "water-louse"; and the Decapoda, or ten-footed animals.

The Entomostraca may be said to include most of the smaller, free-swimming Crustacea, but the word is sometimes used in a stricter and more limited sense. The bodies of the Entomostraca are more or less distinctly jointed, and are contained in a horny, leathery, or brittle shell, formed of one or more parts. The shell is composed of chitin impregnated with a variable amount of carbonate of lime. It is often transparent, and may be striated, reticulated, notched, spinous, etc. It varies in structure in different genera. It may be a bivalve, like a mussel-shell, or folded so as to give the appearance of a bivalve without being really so, or segmented, like a lobster's shell. The body of the organism is segmented, and there is generally a cephalo-thorax region and an abdominal region. In some cases there are distinct head and tail regions. There are one or two pairs of antennæ springing from near the head. The feet vary in number, position, and character. In some genera they are flattened and have branchiæ, or breathing-plates, attached to them, enabling them to perform the function of respiration. There is one conspicuous eye, usually black or reddish, situated in the head region. Near the mouth are two mandibles, and near them are the maxillæ, or foot-jaws, armed with spines or claws and sometimes with branchiæ. There is a heart, often square, that causes the circulation of colorless blood; and well-marked digestive, muscular, nervous, and reproductive systems. The eggs of the Entomostraca may be seen in brood-cavities inside the shell or in exterior attached egg-sacs. The young often hatch in the nauplius form, and undergo several changes before arriving at the adult condition.

The Entomostraca are usually divided into four orders—Copepoda, Ostracoda, Cladocera, and Phyllopoda. The last three are sometimes placed as sub-orders under the order Branchiopoda.

ORDER COPEPODA.

Shell jointed, forming a more or less cylindrical buckler, or carapace, enclosing the head and thorax. The anterior part of the body is composed of ten segments more or less fused. The five constituting the head bear respectively a pair of jointed antennæ, a pair of branched antennules, a pair of mandibles, or masticatory organs, a pair of maxillæ, and a pair of foot-jaws. The five thoracic segments bear **five pairs of jointed swimming-feet**, the fifth often rudimentary. There are about five abdominal segments, nearly devoid of appendages, and continued posteriorly by two tail-like stylets. Young hatched in the nauplius state.

The Copepoda move by vigorous leaps. They lead a roving, predatory life and well deserve the name of "scavengers."

Cyclops.
Copepoda with head hardly distinguishable from the body. The thorax and abdomen generally distinguishable, the former having four and the latter six segments. Two pairs of antennæ, the superior large and many-jointed, the inferior smaller, furnished with short setæ; both superior antennæ of the male have swollen joints. The antennæ assist in locomotion. Two pairs of vigorous branched foot-jaws. One eye, large, single, central. Two egg-sacs. Cyclops are very prolific, as many as 30 or 40 ova being laid at a time and broods occurring at short intervals. The eggs may hatch after leaving the ovary. There are many species. (Pl. XVII, Fig. 5.)

Diaptomus.
Copepoda resembling Cyclops in their general appearance. Thorax and abdomen each five-segmented. Antennæ very long, many-jointed, with setæ; the right antenna only swollen in the male. Antennules large, bifid, the two unequal branches arising from a common footstalk. Three pairs of unbranched foot-jaws. One egg-sac. The ova hatch while borne by the female. (Pl. XVII, Fig. 6.)

Canthocamptus.
Copepoda somewhat resembling Cyclops. The ten segments of the thorax and abdomen not distinguishable. The segments decrease in size as they descend. At the junction of the fourth and fifth segments the body is very movable. Antennæ very short. Five pairs of swimming-feet, much longer than in cyclops. One egg-sac. (Pl. XVII, Fig. 7.)

ORDER OSTRACODA.

Shell consisting of two valves, entirely enclosing the body; from one to three pairs of feet; no external ovary.

Cypris.
Body enclosed within a horny bivalve shell, oval or reniform. Superior antennæ seven-jointed, with long feathery filaments arising from the last three. Inferior antennæ leg-like, with claws and setæ at the end. Two pairs of feet. Eye single. Color greenish, brownish, or whitish. A large number of species. The shell is seldom open wide. (Pl. XVII, Fig. 8.)

ORDER CLADOCERA.

Shell consisting of two thin chitinous plates springing from the maxillary segment. The most important characteristic is the presence of several pairs of leaf-like feet provided with branchiæ, or breathing-organs. There is a large single eye. Two pairs of antennæ, large, branched, and adapted for swimming. This order contains a number of common genera.

Daphnia.
Head produced into a prominent beak; valves of the carapace oval, reticulated, and terminated below by a serrated spine. Superior antennæ situated beneath the beak, one-jointed or as a minute tubercle with a tuft of setæ. Inferior antennæ large and powerful, two-branched, one branch three-jointed, the other four-jointed. Five pairs of legs. Heart a colorless organ at the back of the head. Eye spherical, with numerous lenses. Ova carried in a cavity between the back of the animal and the shell. At certain seasons "winter eggs" are produced. Daphnia move with a louse-like, skipping movement. They are sometimes called "arborescent water-fleas." There are numerous species. (Pl. XVII, Fig. 9.)

Bosmina.
Head terminated in front by a sharp beak directed forwards and downwards, and from the end of which project the long,

many-jointed, curved, and cylindrical superior antennæ. Inferior antennæ two-branched, one branch three-, the other four-jointed. Five pairs of legs. Shell oval, with a spine at the lower angle of the posterior border. Eye large. Eggs hatched in a brood-cavity at the back of the shell. (Pl. XVII, Fig. 10.)

Sida.

Shell long and narrow. Head separated from the body by a depression. Posterior margin nearly straight. No spine or tooth. Antennæ large, one two-jointed, one three-jointed. Six pairs of legs. (Pl. XVIII, Fig. 1.)

Chydorus.

Shell nearly spherical; beak long and sharp, curved downwards and forwards. Antennæ short. Eye single. Color greenish or dark reddish. Moves with an unsteady rolling motion. (Pl. XVIII, Fig. 2.)

ORDER PHYLLOPODA.

Body with or without a shell. Legs 11 to 60 pairs; joints foliaceous or branchiform, chiefly adapted for respiration and not motion. Two or more eyes. One or two pairs of antennæ, neither adapted for swimming.

Branchipus.

Body without a shell. Legs eleven pairs. Antennæ two pairs, the inferior horn-like and with prehensile appendages in the male. Tail formed of two plates. Cephalic horns, with fan-shaped appendages at the base. Color reddish. Floats slowly on its back. (Pl. XVIII, Fig. 3.)

CHAPTER XXII.

BRYOZOA, OR POLYZOA.

The Bryozoa, or Polyozoa, are minute animals forming moss-like or coral-like calcareous or chitinous aggregations. The colonies are called corms, polyzoaria, or cœnœcia. They often attain an enormous size. In the adult stage they lead a sedentary life attached to some submerged object. The animals themselves are small, but easily visible to the naked eye. Some of them are covered with a secreted coating, or sheath, that takes the form of a narrow, brown-colored tube; others are embedded in a mass of jelly. The genera that live in the brown, horny tubes form tree-like growths that often attain considerable length. The branches are sometimes an inch long, and each one is the home of an individual polyzoon, or polypid. The branches, or hollow twigs, are separated from the main stalk by partitions, so that, to a certain extent, each polypid lives a separate existence in its own little case, though each was formed from its next lower neighbor by a process of budding.

The body of the organism is a transparent membranous sac, immersed in the jelly or concealed in the brown opaque sheath. It contains a U-shaped alimentary canal, with a contractile œsophagus, a stomach, and an intestine; a muscular system that permits some motion within the case, and that enables the animal to protrude itself from the case and to extend and contract its tentacles; mesenteries in the form of fibrous bands; an ovary; and a rudimentary nervous system. There is no heart and no blood-vessels of any kind.

The most conspicuous part of the animal is the circlet of ciliated tentacles. They are mounted on a sort of platform, or disc, called a lophophore, at the forward end of the body. This lophophore, with its crown of tentacles, may be protruded from the end of the protective tube at the will of the animal. The tentacles themselves may be expanded, giving a beautiful bell-shaped, flower-like appearance. They are hollow and are covered with fine hair-like cilia. They are muscular and can be bent and straightened at will. By their combined action currents in the water are set up towards the mouth, situated just beneath the lophophore. Minute organisms are thus swept in as food.

The Bryozoa increase by a process of budding which gives rise to the branched stalks. There is also a sexual reproduction. Statoblasts, or winter eggs, form within the body and escape after the death of the animal. They are sometimes formed in such abundance as to form patches of scum upon the surface of a pond. The various forms of these statoblasts assist in the classification of the Bryozoa.

The following are some of the important fresh-water genera. There are many marine forms.

Plumatella.
    Zoary confervoid, brown-colored, branched, tubular, branches distinct. Lophophore crescent-shaped. Tentacles numerous, arranged in a double row. Statoblasts elliptical, with a cellular dark-brown annulus, but no spines. (Pl. XVIII, Fig. 6.)

Fredericella.
    Zoary tubular, branched, brown-colored. Lophophore circular. Tentacles about 24, arranged in a single row. Statoblasts elliptical or subspherical, smooth, no spines, without a cellular annulus. (Pl. XVIII, Fig. 4.)

Paludicella.
Zoary tubular, diffusely branched, having the appearance of brown club-shaped cells joined end to end; apertures lateral, near the broad ends of the cells. Lophophore circular. Tentacles sixteen, arranged in a single row. Statoblasts elliptical, without spines, with a cellular bluish-purple annulus. (Pl. XVIII, Fig. 5.)

Pectinatella.
Zoary massive, gelatinous, fixed. Polypids protruding from orifices arranged irregularly upon the surface. Tentacles numerous. Statoblasts circular, with a single row of double hooks, not forked at the tips, as in Cristatella. Common. (Pl. XVIII, Fig. 7.)

Cristatella.
Zoary a mass of jelly, the polypids arranged on the outside, and the tentacles extended beyond the surface. The jelly-mass is usually long and narrow and has the power of moving slowly, creeping over submerged objects. Tentacles numerous, pectinate upon two arms. Statoblasts circular, with two rows of double hooks having forked tips. Rare.

CHAPTER XXIII.

SPONGIDÆ.

THE fresh-water sponges are not of sufficient importance in water-supplies to warrant an extended description in this work. They differ materially from the marine sponges, which make up by far the greater part of the Spongidæ.

The fresh-water sponge is an agglomeration of animal cells into a gelatinous mass, often referred to as the "sarcode." Embedded in the sarcode and supporting it are minute siliceous needles, or spicules. These skeleton spicules interlace and give the sponge-mass a certain amount of rigidity. The sponge grows as flat patches upon the sides of water-pipes and conduits and upon submerged objects in ponds and streams; or it extends outward in large masses or in finger-like processes that sometimes branch. Its color when exposed to the light is greenish or brownish, but in the dark places of a water-supply system its color is much lighter and is sometimes creamy white. The sponge feeds upon the microscopic organisms in water, which are drawn in through an elaborate system of pores and canals. If these pores become choked up with silt and amorphous matter the organism dies. For this reason sponge-patches are more abundant upon the top and sides of a conduit than upon the bottom.

At certain seasons the fresh-water sponges contain seedlike bodies known under the various names of gemmules, ovaria, statoblasts, statospheres, winter-buds, etc. They are nearly spherical and are about 0.5 mm. in diameter. They have a chitinous coat that encloses a compact mass of protoplasmic globules. In this coat there is a circular orifice, known as the foraminal aperture, through which the protoplasm bodies make their exit at time of germination. In most species the chitinous coat is surrounded by a "crust" in which are embedded minute spicules, called the "gemmule spicules," to distinguish them from the "skeleton spicules," referred to above. There is a third kind of spicule known as the "dermal spicule" or the "flesh spicule." They lie upon the outer lining of the canals in the deeper portions of the sponge. They are smaller than the skeleton spicules and are not bound together. Dermal spicules are not found in all species.

The skeleton spicules differ somewhat in different species. They have a length of about 250 μ. They are usually arcuate and pointed at the ends. They may be smooth or covered with spines (Pl. XVIII, Figs. 9). These skeleton spicules of sponge are commonly observed in the microscopical examination of surface-waters. The gemmule spicules differ in character in different genera and species. Their characteristics are used therefore in classifying the fresh-water sponges.

Potts has described a number of different genera of freshwater Spongidæ, among which are Spongilla, Meyenia, Heteromeyenia, Tubella, Parmula, Carterius, etc. The first two are the most important. They are sometimes given the rank of sub-families.

The Spongilla is a green, branching sponge. The skeleton spicules are smooth and fasciculated. The dermal spicules are fusiform, pointed, and entirely spined. The gemmule spicules are cylindrical, more or less curved, and sparsely spined—the spines often recurved. (Pl. XVIII, Fig. 8.)

The Meyenia are usually sessile and massive. The skeleton spicules are fusiform-acerate, abruptly pointed, coarsely spined except near the extremities; spines subconical, acute. The dermal spicules are generally absent. The gemmule spicules are irregular, birotulate, with rotules produced.

CHAPTER XXIV.

MISCELLANEOUS ORGANISMS.

THE miscellaneous higher animals and plants that one is likely to observe in a microscopical examination of drinking-water are so varied, and they are of such little practical importance in the interpretation of an analysis, that their description here is not warranted. It is sufficient to mention the names of a few common forms.

Of the Vermes the following may be noted: Anguillula, a small, colorless thread-worm like the vinegar-eel (Pl. XIX, Fig. 1); Gordius, the common hair-snake; Nais, an annulate worm with bristles (Pl. XIX, Fig. 2); Tubifex, another bristle-bearing worm; Chætonotus, an elongated worm-like organism with scales on its back (Pl. XIX, Fig. 3). Of the Arachnida: Macrobiotus, the water-bear (Pl. XIX, Fig. 4); and the Acarina, water-mites, or water-spiders (Pl. XIX, Fig. 5). Of the Hydrozoa: the Hydra, a most interesting organism from a zoological standpoint (Pl. XIX, Fig. 6). Insect larvæ; Corethra, or the phantom larva; scales and fragments of insects; barbs of feathers; epithelium-cells; ova of the Entozoa, Crustacea, Rotifera, etc.

Of the vegetable kingdom may be mentioned Batrachospermum (Pl. XIX, Fig. 7); fragments of Sphagnum Moss; Myriophyllum, or water-milfoil; Ceratophyllum, or hornwort (Pl. XIX, Fig. 10); Lemna, or duck-weed (Pl. XIX, Fig. 12); Potamogeton, or pond-weed (Pl. XIX, Fig. 11); Hippuris, or mare's-tail; Anacharis, or American water-weed (Pl. XIX, Fig. 9); Utricularia, an insectivorous plant; pollen-grains; plant-hairs; fragments of vegetable fibres and tissue; fibres of cotton, wool, silk, hemp, etc.; starch-grains, etc.

For the description of all these miscellanous organisms and objects the reader is referred to more comprehensive books on zoology, botany, and general microscopy.

APPENDIX A.

COLLECTION OF SAMPLES.

It cannot be too strongly emphasized that samples of water for analysis must be collected with great care. Whenever possible the analyst himself should supervise the collection. If he attempts to draw inferences from analyses of samples of water about the collection of which he knows nothing he does so at the risk of his reputation.

The quantity of water required for a microscopical examination depends upon the nature of the water. Usually one quart is sufficient, but a gallon is to be preferred and this amount is necessary when a chemical analysis also is to be made. Glass-stoppered bottles should be used, and they should be scrupulously clean. When sent by express they should be packed in covered boxes that have compartments lined with suitable packing-paper to prevent breaking. In winter it may be necessary to use a felt lining to prevent freezing.

If collecting a sample of water from a service-tap, allow the water to run for several minutes before filling the bottle. Rinse the bottle several times before the final filling. Do not fill the bottle completely, but leave a small air-space. If collecting a sample from a stream use care not to stir up the deposit on the bottom, and do not allow floating masses of vegetable matter to enter the bottle. This may be sometimes prevented by pointing the mouth of the bottle down stream. If collecting a sample from a pond use judgment in securing a representative sample. Do not fill the bottle in such a way that the surface-scum may enter. When collecting samples from streams or lakes note carefully the nature of the littoral growths in the vicinity. These are sometimes of value in the interpretation of an analysis.

Numerous methods have been suggested for collecting samples from depths below the surface. The simplest method consists of lowering a weighted stoppered bottle to the desired depth and pulling out the stopper by means of a separate cord. When the bottle is full it may be drawn to the surface with little probability that the water will be displaced. An extra precaution to avoid admixture with the upper layers of water may be taken by using a rubber stopper fitted with a glass tube bent at right angles above the stopper and sealed at the end. With this arrangement the water is allowed to enter the bottle by breaking the glass tube by a pull from an auxiliary cord. Or an inflated rubber ball may be put into the bottle. When the water enters, the ball will be forced up into the neck of the bottle on the inside and make an effective seal.

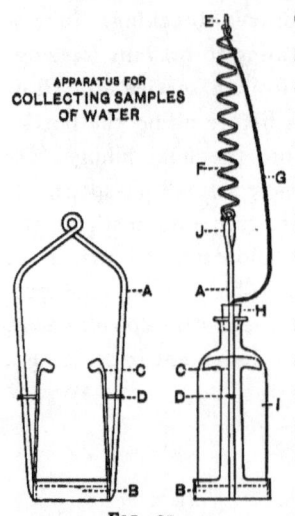

APPARATUS FOR COLLECTING SAMPLES OF WATER

FIG. 21.

When collecting samples from depths greater than 50 ft. it is desirable to avoid the use of the auxiliary cord. The following apparatus has proved very satisfactory down to depths of 400 ft. (See Fig. 21.)

The frame for holding the bottle consists of a brass wire, A, attached to a weight, B, which is made by rolling a sheet of brass so as to form the sides of a shallow pan and filling this with melted lead to the height indicated by the dotted line. At each side where the wire rod is attached a strip of brass extends upward, terminating in a clip, C. These brass strips have considerable spring and are designed to hold the bottle in place, as shown in the cut. Guides, D, prevent the strips from being bent too far inward, and the uprights, A, prevent them from being bent too far outward. The bottle may be inserted easily by holding back the springs, C, and pushing it between the clips. The frame is supported by the spring, F, joined to the sinking-rope, E. A flexible cord, G, extends from the top of the spring, E, to the stopper, H, of the bottle, I. The length of this cord and the length and stiffness of the spring are so adjusted that when the apparatus is suspended in the water by the sinking-rope the cord will be just a little slack. In this condition it is lowered to the depth at which one wishes to fill the bottle. A sudden jerk given to the rope stretches the spring and produces sufficient tension on the cord, G, to pull out the stopper. As a precaution against a possible loss of the apparatus through breaking of the spring, a safety-cord, not shown in the figure, extends through the helix connecting the sinking-rope, E, directly to the frame, J. This safety-cord, which is always somewhat slack, is also adjusted to prevent too great a stretching of the spring.

With great depths it is necessary to reduce the size of the aperture through which the water enters the bottle and to close this with a suitable valve. This may be done by passing a piece of brass tube through a rubber stopper and closing this tube at the top with a brass plug ground to fit. Or the spring may be used to break the end of a sealed glass tube inserted in the stopper.

APPENDIX B.

TABLES AND FORMULÆ.

WEIGHTS AND MEASURES—CONVERSION TABLES.

1 lb. Avoir. = 1.215 lbs. Troy or Apoth. = 7000 grains Troy = 453.6 grams.
1 lb. Troy or Apoth. = .823 lb. Avoir. = 5760 grains Troy = 373.2 grams.
1 oz. Avoir. = .960 fluid ounce = 28.35 grams.
1 oz. Troy or Apoth. = 1.053 fluid ounces = 31.10 grams.
1 grain Troy = .0648 gram.
1 kilogram = 2.205 lbs. Avoir. = 2.679 lbs. Troy or Apoth.
1 gram = .035 oz. Avoir. = .032 oz. Troy or Apoth. = 15.432 grains Troy.
1 milligram = .0154 grain Troy.
1 Imperial gallon = 1.201 U. S. fluid gallons = 277.4 cubic inches = 4546 cubic centimeters.
1 U. S. fluid gallon = .833 Imperial gallon = 231 cubic inches = 3785 cubic centimeters.
1 U. S. fluid gallon = 8.332 lbs. Avoir. = 10.127 lbs. Troy or Apoth.
1 fluid ounce = 1.042 oz. Avoir. = .949 oz. Troy or Apoth. = 29.57 cubic centimeters.
1 liter = .264 U. S. fluid gallon = .220 Imperial gallon = 21.028 cubic inches.
1 liter = 33.82 fluid ounces = 2.205 lbs. Avoir. = 2.679 lbs. Troy or Apoth.
1 cubic centimeter = .033 fluid ounce = .035 oz. Avoir. = .032 oz. Troy or Apoth.

1 inch = 2.54 centimeters = 25.4 millimeters.
1 foot = 30.48 centimeters.
1 yard = 91.44 centimeters = .9144 meter.
1 meter = 1.0936 yards = 3.28 feet = 39.37 inches.
1 centimeter = .3937 inch.
1 millimeter = .0394 inch = .442 Paris lines.
1 micron (μ) = .001 millimeter = $\frac{1}{25400}$ inch = .000039 inch = .0004 Paris line.
1 Paris line = .089 inch = 2.26 millimeters = 2260.6 microns.

1 cubic yard = .7645 cubic meter.
1 cubic foot = .0283 cubic meter = 7.481 U. S. gallons = 6.232 Imperial gallons.
1 cubic inch = 16.39 cubic centimeters.
1 cubic meter = 35.216 cubic feet = 1.308 cubic yards.
1 cubic centimeter = .061 cubic inch.

LABORATORY TABLES AND FORMULÆ.

TABLE FOR TRANSFORMING MICROMILLIMETERS (MICRONS) TO INCHES.

Microns.	Decimals of an Inch.	Fractions of an Inch.	Microns.	Decimals of an Inch.	Fractions of an Inch.
1	.000039	1/25000	25	.000984	1/1000
2	.000079	1/12500	30	.001181	1/833
3	.000118	1/8333	35	.001378	1/714
4	.000157	1/6250	40	.001575	1/625
5	.000197	1/5000	45	.001772	1/533
6	.000236	1/4333	50	.001969	1/500
7	.000276	1/3285	60	.002362	1/416
8	.000315	1/3125	70	.002756	1/357
9	.000354	1/2777	80	.003150	1/312
10	.000394	1/2500	90	.003543	1/277
15	.000591	1/1666	100	.003937	1/250
20	.000787	1/1250			

TABLE FOR TRANSFORMING CENTIGRADE TO FAHRENHEIT DEGREES OF TEMPERATURE.

Centigrade.	Fahrenheit.	Centigrade.	Fahrenheit.	Centigrade.	Fahrenheit.
−17.7	0	4.0	39.2	23.8	75.0
−15.0	5.0	4.4	40.0	25.0	77.0
−12.2	10.0	5.0	41.0	26.6	80.0
−10.0	14.0	7.2	45.0	29.4	85.0
−9.4	15.0	10.0	50.0	30.0	86.0
−6.6	20.0	12.7	55.0	32.2	90.0
−5.0	23.0	15.0	59.0	35.0	95.0
−3.8	25.0	15.5	60.0	37.7	100.0
−1.1	30.0	18.3	65.0	40.0	104.0
0	32.0	20.0	68.0		
1.6	35.0	21.1	70.0		

TABLE FOR TRANSFORMING STATEMENTS OF CHEMICAL COMPOSITION.

	Grains per U. S. Gallon.	Grains per Imp. Gallon.	Parts per 100,000.	Parts per 1,000,000.
1 grain per U. S. gallon....	1.	1.20	1.71	17.1
1 grain per Imperial gallon.	0.830	1.	1.43	14.3
1 part per 100,000..........	0.585	0.70	1.	10.0
1 part per 1,000,000........	0.058	0.07	0.10	1.

APPENDIX B.

TABLE* FOR TRANSFORMING COLOR-READINGS OF THE NESSLER (NATURAL WATER) SCALE TO THOSE OF THE PLATINUM-COBALT SCALE.

Nessler Scale.	Platinum-cobalt Scale.	Nessler Scale.	Platinum-cobalt Scale.	Nessler Scale.	Platinum-cobalt Scale.
0	0	.70	.58	1.42	1.10
.06	.10	.74	.60	1.50	1.16
.10	.18	.80	.63	1.56	1.20
.13	.20	.90	.70	1.60	1.22
.20	.26	.99	.80	1.70	1.29
.26	.30	1.00	.81	1.72	1.30
.30	.33	1.10	.88	1.80	1.36
.40	.39	1.13	.90	1.86	1.40
.42	.40	1.20	.95	1.90	1.43
.50	.46	1.27	1.00	2.00	1.50
.57	.50	1.30	1.02		
.60	.52	1.40	1.09		

* Based upon several series of comparisons by the analysts of the Boston Water Supply Department.

DIRECTIONS FOR CLEANING GLASSWARE.

To clean bottles to be used for collecting samples of water.—Wash with chromic acid prepared by saturating strong sulphuric acid with potassium bichromate. Rinse thoroughly several times with distilled water. Drain and dry. To remove the gelatinous film that sometimes collects, use shot, clean gravel, or cotton waste and sand, and afterwards wash with acid.

To clean cover-slips.—Immerse for a few hours, or boil, in nitric acid, or in chromic acid prepared as above. Rinse in water, and store in alcohol to which a little ammonia has been added.

To clean counting-cells.—Wash with cold distilled water and wipe dry with a clean linen cloth free from lint. By blowing a stream of water from a wash-bottle into the corners of the cell the organisms may be prevented from becoming lodged there.

PRESERVATION OF MICROSCOPIC ORGANISMS.

The microscopic organisms may be preserved in permanent mounts upon glass slips according to methods described in the various text-books on microscopical technique. For practical study it is more convenient to preserve them in mass in 2-oz. bottles. For this purpose the following killing and preservative fluids may be found useful:

King's Fluid (for preserving algæ, etc.).—

Camphor-water*	50 grams.
Distilled water	50 "
Glacial acetic acid	0.50 "
Copper nitrate, crystals	0.20 "
Copper chloride, crystals	0.20 "

Corrosive Acetic Acid (for killing).—Saturated solution of mercuric chloride plus 10% of acetic acid. After using, wash with water. Preserve in alcohol.

Formaldehyde.—For killing, use a 40% solution, sold under the name of "Formalin." For preserving, use solutions varying from 5% to 10%, according to the organisms.

Picro-sulphuric Acid (for killing).—

Distilled water saturated with picric acid	100 c.c.
Sulphuric acid, strong	2 c.c.

After using, wash with 60% alcohol.

Corrosive Sublimate (for killing Protozoa).—To water containing the organisms add an equal volume of saturated corrosive sublimate. Decant, and add 50% alcohol, changing this in an hour to 70%.

---

* Made by letting a lump of camphor stand in distilled water for a few days.

APPENDIX C.

BIBLIOGRAPHY.

The following is a partial list of references to articles on the microscopic organisms and their relation to drinking-water, together with such other references as will enable the student to investigate the broader subjects of sanitary water-analysis and limnology.

MICROSCOPY.

Bausch, Edw. Manipulation of the Microscope. Rochester, N. Y.: Bausch & Lomb Optical Co.
Beale, L. S. How to Work with the Microscope. 5th edition. Philadelphia: Lindsay & Blakiston, 1880.
Behrens, J. W. A Guide for the Microscopical Investigation of Vegetable Substances. Translated by Rev. A. B. Hervey. Boston: S. E. Cassino, 1885.
Carpenter, W. B. The Microscope and its Revelations. 7th edition. Edited by Dallinger. Philadelphia: P. Blakiston, Son & Co., 1891.
Davis, Geo. E. Practical Microscopy. 3d edition. Philadelphia: J. B. Lippincott Co., 1889.
Davis and Mathews. The Preparation and Mounting of Microscopic Objects. New York: G. P. Putnam's Sons, 1890.
Deby, Julian. A Bibliography of the Microscope and Micrographic Studies. London: D. Bogue, 1882.
Frey, H. The Microscope and Microscopical Technology. New York: Wm. Wood & Co., 1880.
Gage, S. H. The Microscope and Microscopical Methods. 7th edition. Ithaca, N. Y.: Comstock Pub. Co.

Lankester, E. Half-hours with the Microscope: A Popular Guide to the Use of the Microscope as a Means of Amusement and Instruction. 20th edition. London, 1898.
Naegeli and Schwendener. The Microscope in Theory and Practice. 2d edition. London: Swan, Sonnenschein, Lowry & Co.
Nave, J. Collector's Handy-book. London: W. H. Allen & Co.
Pringle, Andrew. Practical Photo-micrography. New York: Scovell & Adams Co., 1890.
Van Heurck, H. Le Microscope—sa construction, son mainiement, et son application speciale à l'anatomie végétale et aux diatomées. 3d edition. Brussels, 1878.

BIOLOGY, BOTANY, ZOOLOGY.

Bessey, C. F. Botany (Advanced Course). New York: Henry Holt & Co., 1888.
Davenport, Chas. B. Experimental Morphology. Part I. Effect of Chemical and Physical Agents upon Protoplasm. New York: Macmillan Co. 1897
Dragendorff, G. Plant Analysis, Qualitative and Quantitative.
Goebel, K. Outlines of Classification and Special Morphology of Plants.
Huxley and Martin. A Course of Elementary Instruction in Practical Biology. London: Macmillan & Co., 1883.
Klebs, G. Die Bedingungen d. Fortpflanzung bei einigen Algen u. Pilzen. Jena: Gustav Fischer, 1896.
Mandel, John A. Hand-book for Bio-chemical Laboratory. New York: John Wiley & Sons, 1896.
Parker, T. Jeffrey and Haswell, W. A. A Text-book of Zoology. London: Macmillan & Co., 1897.
Poulsen, V. A. Botanical Micro-chemistry. Translated by Wm. Trelease. Boston: Cassino & Co., 1884.
Ranvier. Traité d'Histologie. Paris: Savy, 1875, 1882.
Sachs, Julius. Text-book of Botany. 2d edition. Oxford: Clarendon Press, 1882.
Schäfer. Essentials of Histology. Philadelphia: Lea, 1885.
Sedgwick, Wm. T., and E. B. Wilson. General Biology. New York: Henry Holt & Co., 1895.
Stöhr. Text-book of Histology. Translation by Schafer. Philadelphia: Blakiston, 1896.
Strasburger, E. Microscopic Botany: A Manual of the Microscope in Vegetable Histology. Boston: S. E. Cassino, 1887.

Taylor, J. E. The Aquarium : Its Inhabitants, Structure, and Management. London : W. H. Allen & Co., 1884.
Thomson, J. A. The Study of Animal Life. University Extension Manuals. New York, 1892.
Vines. Students' Text-book of Botany. London : Sonnenschein, 1894, 1895.
Warming. Text-book of Botany. Translation by M. C. Potter. London : Sonnenschein, 1895.
Zimmermann, A. Die botanische Mikrotechnik. Ein Handbuch der mikroskopischen Präparations-, Reactions-, und Tinctionsmethoden. Tübingen, 1892.

MICROSCOPICAL EXAMINATION OF WATER.

Bell, James. Microscopical Examination of Water for Domestic Use. Mo. Micro. Jour., V, 163. London, 1871.
Calkins, Gary N. The Microscopical Examination of Water. 23d An. Rept. Mass. St. Bd. of Health, 1891.
Certes, A. Analyse Micrographique des Eaux. Paris : B. Tignol, 1883.
Drown, Thomas M. The Analysis of Water : Chemical, Microscopical, and Bacteriological. Jour. N. E. Water Works Assoc., Dec., 1889.
Hansen, S. Ch. Methode zur Analyse des Brauwassers in Rücksicht auf Mikroorganismen. Centb. f. Bacter. et Parasitenk., III, 1888.
Harz, C. O. Mikroskopische Untersuchung des Brunnenwassers für hygienische Zwecke. Zeitschrift f. Biologie, XII, 100, 1876.
Hassall, A. H. A Microscopic Examination of the Water Supplied to the Inhabitants of London and the Suburban Districts. London, 1850.
———— Food : Its Adulterations and the Methods for their Detection. London : Longmans, Green & Co., 1876.
Hirt, L. Ueber den Principien und die Methode der Mikroskopischen Untersuchung des Wassers. Zeitschrift für Biologie, 1879.
Hitchcock, R. The Biological Examination of Water. Am. Mo. Micr. Jour., VIII, 9, 1887.
Hovenden, F. Examining Thin Films of Water. 18th An. Rept. London Micro. and Nat. Hist. Club, 1889, 10.
Hulwa, Franz. Beiträge zur Schwemmkanalization und Wasser-Versorgung der Stadt Breslau. Centralblatt für allgemeine Gesundheitspflege, Ergänzungsheft, I, 89. Bronn, 1885.
Jackson, D. D. On an Improvement in the Sedgwick-Rafter Method for the Microscopical Examination of Drinking Water. Tech. Quarterly, IX, Dec., 1896.

Jackson, D. D. An Improved Filter for Microscopical Water Analysis. Tech. Quarterly, XI, Dec., 1898.
Jolles, M. Die bacteriologische und mikroskopische Wasseruntersuchung. Wien, 1892.
Kean, A. L. A New Method for the Microscopical Examination of Water. Science, Feb. 15, 1889; Eng. News, March 30, 1889.
Leeds, A. R. Quantitative Estimation of Micro-Organisms. The Stevens Indicator, XIV, Jan., 1897.
MacDonald, J. D. A Guide to the Microscopical Examination of Drinking Water. 2d edition. London: J. & A. Churchill, 1883.
Mez, Carl. Mikroskopische Wasseranalyse. Berlin: Julius Springer, 1898.
Neuvelle. Des Eaux de Paris: Essai d'analyse micrographique comparée. Paris, 1880.
Radlkofer, L. Mikroskopische Untersuchung der organischen Substanzen im Brunnenwasser. Zeitschrift für Biologie, 1865.
Rafter, George W. On the Use of the Microscope in Determining the Sanitary Value of Potable Water, with Special Reference to the Study of the Biology of the Water of Hemlock Lake. Proc. Micro. Sect. Rochester Acad. of Sciences, 1886.
——— How to Study the Biology of a Water Supply. A paper read before the Section of Microscopy, Rochester Acad. of Sciences.
——— The Biological Examination of Potable Water. Proc. Rochester Acad. Sciences, I, 33–44. Rochester, 1890.
——— On Some Recent Advances in Water Analysis and the Use of the Microscope for the Detection of Sewage Contamination. Am. Month. Micro. Jour., May, 1893.
——— The Microscopical Examination of Potable Water. No. 103 in Van Nostrand Science Series.
Sedgwick, William T. Biological Examination of Water, Technology Quarterly, II., 67.
——— Biological Water Analysis. Proc. Soc. Arts, 1888–89.
——— Recent Progress in Biological Water Analysis. Jour. N. E. W. W. Assoc., IV, Sept., 1889.
Sorby, H. C. Microscopical Examination of Water for Organic Impurities. Jour. Roy. Micro. Sci., Series 2, IV, 1884.
——— Detection of Sewage Contamination by the Use of the Microscope and on the Purifying Action of Minute Animals and Plants. Jour. Soc. Arts, XXXII, 929, 1884; Jour. Roy. Micr. Soc., Series 2, IV, 988, 1884.
Tate, A. N. Microscopical Examination of Potable Water. Engl. Mechanic, XLI, 1885.

Thomé, O. W. Zur mikroskopischen Untersuchung von Brunnenwasser. Zeitschrift für Biologie, III, 258.
Whipple, Geo. C. A Standard Unit of Size for Micro-Organisms. Am-Monthly Micro. Jour., XV, Dec., 1894.
—— Experience with the Sedgwick-Rafter Method. Technology Quarterly, IX, Dec., 1896.
—— Microscopical Examination of Water, with a Description of a Simple Form of Apparatus. Science, VI, July 16, 1897.
Wolff, A. J. The Sanitary Examination of Drinking Water. 8th An. Rept. Conn. St. Bd. of Health, 1885.
Zune, A. J. Traité d'Analyse Chimique, Micrographique et Microbiologique des Eaux Potables. Paris, 1894.

PHYSICAL AND CHEMICAL EXAMINATION OF WATER.

Austin, G. L. Water Analysis: A Hand-book for Water Drinkers. New York: C. T. Dillingham, 1883.
Davis, Prof. Floyd. Interpretation of Water Analyses. Municipal Engineering, XIV, June, 1896.
Drown, Thomas M. Chemical Examination of Drinking Water. Proc. Soc. Arts, 1887–88.
—— The Odor and Color of Surface Waters. Jour. N. E. Water Works Assoc., March, 1888.
FitzGerald and Foss. On the Color of Water, with Description of Colorimeter. Jour. Frank. Inst., CXXXVIII, Dec., 1894.
Frankland, E. Water Analysis for Sanitary Purposes. Philadelphia, 1880.
Hazen, Allen. A New Color-standard for Natural Waters. Am. Chem. Jour., XIV.
Hollis, F. S. An Investigation of the Cause of the Color of Natural Water. Rept. of Boston Water Board for 1892, 95–115.
—— Methods for the Determination of Color of Water and the Relation of Color to the Character of Water. Jour. N. E. W. W. Assoc., 1898.
Hoppe-Seyler. Ueber die Verteilung absorbierter Gase im Wasser des Bodensees und ihre Beziehungen zu den in ihm lebenden Thieren und Pflanzen. In Schriften des Vereins f. Gesch. d. Bodensees u. s. Umgebung, Heft 24, 1895.
Hornung, George. Diaphanometer for Measuring the Transparency of Water. A paper read before the Engineers' Club of Cincinnati, March 19, 1896.

Kemna, Ad. La Couleur des Eaux. Bull. d. l. Société Belge de Géologie, de Paléontologie et Hydrologie (Bruxelles), X, 224–279, 1896.
Leffmann and Bean. Examination of Water for Sanitary and Technical Purposes. 3d edition. Philadelphia: P. Blakiston & Co., 1895.
Lovibond, Jos. W. Patent Tintometer: A Descriptive Catalogue.
Mallet, J. W. Reports (1, 2, 3) on the Results of an Investigation as to the Chemical Methods for the Determination of Organic Matter in Potable Water. An. Rept. Nat. Bd. Health for year ending June 30, 1882, 184–354.
Nichols, Wm. Ripley. Water Supply Considered Mainly from a Chemical and Sanitary Standpoint. New York: John Wiley & Sons, 1883.
Pearmain, T. H., and C. G. Moor. The Chemical and Biological Analysis of Water. A series of papers in the Sanitary Record for 1898.
Richards, Ellen H. Laboratory Notes on Chemistry and Water Analysis. Pamphlet (not published) printed for the use of students of the Mass. Inst. of Technology.
Richards, Ellen H., and J. W. Ellms. The Coloring Matter of Natural Waters, its Source, Composition, and Quantitative Measurement. Jour. Am. Chem. Soc., XVII, Jan., 1896.
Rideal, Samuel. Water and its Purification. Philadelphia: Lippincott, 1897.
Seddon, James A. Clearing Water by Settlement: Observations and Theory. Jour. of Assoc. of Eng. Soc., VIII, Oct., 1889.
Tiemann and Gärtner. Die chemische und mikroskopische-bakteriologische Untersuchung des Wassers. Braunschweig: F. Vieveg & Son, 1889.
Thorpe's Dictionary of Applied Chemistry, Art. Water.
Wanklyn, J. Alfred. Water Analysis: A Practical Treatise on the Examination of Potable Water. London: Trübner & Co., 1884.
Ziegeler, G. A. Die Analyse des Wassers. Stuttgart (Enke), 1887.

BACTERIOLOGICAL EXAMINATION OF WATER.

Abbott, A. C. The Principles of Bacteriology. Philadelphia: Lea Brothers & Co., 1895.
Crookshank, Edgar M. Manual of Bacteriology. London, 1890.
Frankland, Percy and G. C. Micro-organisms in Water. London, 1894.
Fuller, G. W., and W. R. Copeland. Quantitative Determination of Bacteria in Sewage and Water. An. Rept. Mass. St. Bd. of Health. 1895.

Lehmann, K. B., and Rudolph Neumann. Atlas and Essentials of Bacteriology. München: Lehmann, 1896.

Muir and Ritchie. Manual of Bacteriology. London, 1897.

Papers and Proceedings of the Convention of Bacteriologists, held at New York June 21–22, 1895, under the auspices of the Committee on the Pollution of Water Supplies of the American Public Health Association. Jour. Am. Public Health Assoc., Oct., 1895.

Pearmain and Moor. Applied Bacteriology. London: Baillière, 1897.

Procedures Recommended for the Study of Bacteria. The Report of a Committee of American Bacteriologists to the Committee on Pollution of Water Supplies of the American Public Health Association. Jour. Am. Pub. Health Assoc., 1898.

Smith, Theobald. A New Method for Determining Quantitatively the Pollution of Water by Fecal Bacteria. 13th An. Rept. N. Y. St. Bd. of Health, 713, 1893.

Sternberg, George M. A Manual of Bacteriology. New York: Wm. Wood, 1892.

Woodhead, G. S. Bacteria and their Products. New York, 1891.

LIMNOLOGY.

TEMPERATURE OF WATER.

Agassiz, Alexander. Hydrographical Sketch of Lake Titicaca. Proc. Am. Acad. of Arts and Sci., XI, 283, 1876.

Bachmann, Hans. Mitteilungen der Natursforschenden Gesellschaft. Luzern.

Buchanan, J. Y. Temperature of Lakes. Nature, March 6, 1879.

FitzGerald, Desmond. The Temperature of Lakes. Trans. Am. Soc. Civ. Eng., XXXIV, Aug., 1895.

Ganong, W. F. Upon Temperature-Measurements with the Thermophone in Clear Lake, Lebreau. Bull. No. 14, New Brunswick Nat. Hist. Soc.

Marsh, C. Dwight. Notes on the Depth and Temperature of Green Lake. Trans. Wisconsin Acad. of Sci., Art, and Letters, VIII.

Nichols, William R. On the Temperature of Fresh Water Lakes and Ponds. Proc. Bost. Soc. Nat. Hist., XXI, 1880. [Contains a bibliography of the subject to date.]

Richter, E. Temperaturverhältnisse der Alpenseen. Verh. d. 9. Deutsch. Geographentages zu Wien, 1891.

Russell, Israel C. Lakes of North America. Boston: Ginn & Co., 1895.

de Saussure, H.-B. Voyages dans les Alpes. I, 23. Neuchâtel, 1799.

Smith, Hamilton. Temperature of Lakes. Trans. Am. Soc. Civ. Eng., March, 1884.

Stearns, Frederick P. Temperature of Water. Special Report of Mass. St. Bd. of Health on Examination of Water Supplies, 1890, 659.

Ule. Die Temperaturverhältnisse der Baltischen Seen. Verhandlung d. 10. Deutschen Geographentages in Stuttgart, 1893.

Warren, H. E., and G. C. Whipple. The Thermophone, a new instrument for obtaining the temperature of a distant or inaccessible place; and Some Observations on the Temperature of Surface Waters. Am. Meteor. Jour., XII, June, 1895.

———— The Thermophone, a new instrument for determining temperatures. Technology Quarterly, VIII, July, 1895.

Whipple, Geo. C. Some Observations on the Temperature of Surface-waters, and the Effect of Temperature on the Growth of Microorganisms. Jour. N. E. W. W. Assoc., IX.

———— The Thermophone. Science, Nov. 15, 1895.

———— Classification of Lakes according to Temperature. The American Naturalist, XXXII, Jan., 1898.

TRANSMISSION OF LIGHT BY WATER.

Fol et Sarasin. Pénétration de la Lumière du jour dans les Lacs du Lac de Genève et dans celles de la Méditerranée. Mémoires de la Société de Physique et d'histoire Nat. de Genève, XXIX, 1887.

Forel, F. A. Le Leman, une Monographie Limnologique. 3 Parts.

Linsbauer. Vorschlag einer verbesserten Methode zur Bestimmung der Lichtverhältnisse im Wasser. In Verh. d. k. k. zool-bot. Gesellschaft in Wien, Jahrgang, 1895.

Secchi, Père A. Relazione dell' esperienze fatte a bordo della pontificia pirocorvetta l'Immacolata Concezione per determinare la transparenza del mare. In A. Cialti, Sul moto ondoso del mare. Roma, 1866.

Wild, H. Ueber die Lichtabsorption der Luft. Poggendorfs Annalen. Anhang CXXXIV., 582. Berlin, 1868.

MICROSCOPIC ORGANISMS.

Bailey, J. W. Infusoria in Hudson River Ice. Am. J. Sc., Series 2, XI, 351. New Haven, 1851.

———— Microscopical Examination of Soundings made by the U. S. Coast Survey of Atlantic Coast of U. S. Smith. Contributions to Knowledge, II, Art. 3. Washington, 1860.

Bailey, J. W. On the Infusoria and other Microscopic Forms in Dust Showers and Blood Rain. By C. G. Ehrenberg. Translated by J. W. Bailey. Am. J. Sc. New Haven, 1851.

Bennett and Murray. A Handbook of Cryptogamic Botany. New York: Longmans, Green & Co., 1889.

Campbell, Douglas H. Plants of the Detroit River. Bull. Torrey Bot. Club, 83. New York, 1886.

Collingwood, C. On the Microscopic Algæ which Cause the Coloration of the Sea in Various Parts of the World. Trans. Micro. Soc., XVI, 85. London, 1868.

Cooke, M. C. One Thousand Objects for the Microscope. London: F. Warne & Co.

—— Ponds and Ditches. London, 1885.

Cutter, E. On the Presence of Forms of Life in the Central and Lateral Ground Waters. Am. Mo. Mic. Jour., I, 186. New York, 1880.

Engler, A., and K. Prantl. Die naturlichen Pflanzen-familien. Teil I. Leipzig, 1896.

Eyferth, B. Die einfachsten Lebensformen des Thier- und Pflanzenreiches. Braunschweig, 1885.

Fryer, A. Potamogetons (Pond Weeds) of the British Isles. Parts 1–3. London, 1898.

Griffith and Henfrey. The Micrographic Dictionary, 4th edition. London: J. Van Voorst, 1883.

Hooke, R. Micrographia, or Some Physiological Descriptions of Minute Bodies made by Magnifying Glasses, with Observations and Inquiries Thereupon. London, 1667.

Hoole, Samuel. The Select Works of Antony Van Leeuwenhoek, containing his Microscopical Discoveries in many of the Works of Nature. Translated from the Dutch and Latin editions published by the author. In two volumes. London, 1798.

Kirchner und Blochmann. Die mikroskopische Pflanzen- und Thierwelt des Süsswassers. I. Plants. II. Animals. 2d edition. Hamburg, 1891.

Leeuwenhoek, A. Van. Select Works. Translated by Samuel Hoole. London: G. Sidney, 1800.

Pritchard, Andrew. A History of Infusoria, including the Desmidiaceæ and Diatomaceæ, British and Foreign. London: Whittaker & Co., 1861.

—— Microscopic Illustrations. 3d edition. London: Whittaker & Co., 1845.

Schneidemühl, G. Die Protozoen als Krankheitserreger des Menschen und der Hausthiere. Leipzig: W. Engelmann, 1898.

Stokes, A. C. Aquatic Microscopy. Philadelphia: Queen & Co.

Vorce, C. M. Some Observations on the Minute Forms of Life in the Waters of the Lakes. Cleveland, O., 1880.
—— Microscopic Forms Observed in Water of Lake Erie. Proc. Am. Soc. Micro. for the years 1881 and 1882.
—— Forms Observed in the Water of Lake Erie. Proc. Am. Soc. Micro. Buffalo, 1887.
Wood, J. G. Common Objects for the Microscope. London: G. Routledge & Son.
Zacharias, Otto. Die Thier- und Pflanzenwelt des Süsswassers. 2 Bde., Leipzig, 1891.

DIATOMACEÆ.

Attwood. Diatoms from the Chicago Water Supply. Mo. Micro. Jour. XVII, 266. London.
Blake, Dr. Diatoms found in Hot Springs. Calif. Acad. of Sci. and Mo. Micro. Jour., IX, 71; A. M. N. H., II, ser. 4. London, 1872.
Briggs, S. A. The Diatomaceæ of Lake Michigan. The Lens, I, 41. Chicago, 1872.
Carter, Fredk. B. Diatoms: Their Life History and their Classification. Micro. Jour. Washington, 1891.
Castracane, Conte Abbé F. Reproduction and Multiplication of Diatoms. J. R. M. S., 1889, 22. London, 1889.
—— Reproduction of Diatoms. Ann. de Micrographie, IX, Dec., 1897.
Deby, Julien. A Bibliography of the Diatomaceæ. "A Bibliography of the Microscope," III. London, 1882. Also, Bibliographie Diatomologique. Jour. Microg., XI, 217. Paris, 1887.
Donkin, Arthur S. The Natural History of the British Diatomaceæ. London: J. Van Voorst, 1870.
Forel, F. A. Diatomées du Lac Leman. Lausanne, 1874.
Kitton, Fred. Notes on New York Diatoms, with Descriptions of Fragilaria Crotonensis. K. Sc. Gossip, V, 109, 1869.
Kuetzing, F. Die Bacillarien, oder Diatomeen. Nordhausen, 1844.
Lockwood, S. Raising Diatoms in the Laboratory. Jour. N. Y. Micro. Soc.. II, 153. New York, 1886. J. R. M. S., 1887, 626.
Mills, Fredk. Wm. An Introduction to the Study of the Diatomaceæ. London, 1893; also, The Microscopical Pub. Co., Washington, D. C. Contains an extensive bibliography on the Diatomaceæ by Julien Deby.
Miquel, P. De la culture artificielle des Diatomées. Le Diat., I, 73, 93, 121, 123, 149, 165. Paris.

Miquel, P. Biologie des Diatomées. Ann. de Micr., IV, 321, 1892. T. R. M. S. 1892, 655.
Pelletan, J. Les Diatomées. Paris, 1891.
Petit, Paul. Life History of the Diatomaceæ. J. R. M. S., II, 181. London, 1879.
Schmidt, Adolf. Atlas der Diatomaceen-Kunde. In Verbindung mit den Herren Grundler, Grunow, Janisch, Weissflog und Witt. 1875. (There is a blue-print reproduction of these plates by C. Henry Kain, Camden, N. J.)
Smith, H. L. Conspectus of the Families and the Genera of the Diatomaceæ. Lens, I, 1, 72, 154. Chicago, 1872. Notice in Amer. Naturalist, VI, 318. Salem, 1872.
────── Contribution to the Life History of the Diatomaceæ. Part I. Proc. Amer. Soc. Micr., 1886-87; also, Contribution à l'Histoire Naturelle des Diatomées; also, Journ. de Microg., XII, 22, 507; XIII, 21, 49, 84, 120, 308. Paris, 1888-9.
────── A Contribution to the Life History of the Diatomaceæ. Part II. Proc. Amer. Soc. Micr., 10th Ann. Meet. at Pittsburgh, Pa., Peoria, Ill., 1888.
Smith, Wm. Synopsis of the British Diatomaceæ. 2 vols. London, 1853-56.
Thomas, B. W., and Chase, H. H. Diatomaceæ of Lake Michigan, as collected during the last 16 years from the Water Supply of the City of Chicago. Presented to the State Nat. Hist. Soc. of Illinois, May 14, 1886. Chicago, 1886. Notar. Ann., II, 328. Venezia, 1887.
Van Heurck, H. Synopses des Diatomées de Belgique. Anvers, 1885.
Wolle, Francis. Diatomaceæ of North America. Bethlehem, Pa., 1890.

SCHIZOMYCETES AND FUNGI.

Cohn, F. Ueber den Brunnenfaden, mit Bemerkungen über die mikroskopische Analyse des Brunnenwassers. Beiträge zur Biologie, I, 117-131. 1870.
Cooke and Berkeley. Fungi: Their Nature, Influence, and Uses. 3d edition. London : Kegan Paul, Trench & Co., 1883.
Cooke, M. C. Rust, Smut, Mildew, and Mould: An Introduction to the Study of the Microscopic Fungi. 5th edition. London : W. H. Allen & Co., 1886.

De Bary, A. Comparative Morphology and Biology of the Fungi, Mycetozoa, and Bacteria. Oxford: Clarendon Press, 1887.
Duclaux, M. Chimie Biologique. No. 9 in Encyclopédie Chimique. Paris, 1887.
Jelliffe, Smith Ely. On Some Laboratory Moulds. Jour. of Pharmacology, IV, Nov., 1897.
Lindstedt, R. Synopsis der Saprolegniaceen. 1872.
Nägeli, C. Die niederen Pilze. 1877.
Winogradsky, S. Ueber Eisenbacterien. Botanische Zeitung, XLVI, 1888.
Zopf, Dr. Wilhelm. Die Pilze. Breslau, 1890.
———— Untersuchungen über Crenothrix polyspora. 1879.
———— Die Spaltpilze. Nach dem neuesten Standpunkte bearbeitet, 1883.

CHLOROPHYCEÆ AND CYANOPHYCEÆ.

Agardh, J. G. Species, genera et ordines Algarum. 1848-1863.
Bennett, A. W. Fresh-water Algæ (including Chlorophyllaceous Protophyta) of the English Lake District, with descriptions of twelve new Species. J. R. M. S., Series 2, VI, 1. London, 1886.
Bornet, E., and Ch. Flahault. Tableau synoptique des Nostochacées filamenteuses. Mém. de la Soc. des Sc. Nat. de Cherbourg, XXVI, 1889.
Cohn, F. Untersuchungen über die Entwickelung der Mikroskopischen Algen und Pilze. 1853. Beiträge su: Biologie der Pflanzen. 1870.
Cooke, M. C. Introduction to Fresh-water Algæ, with an enumeration of all the British Species. Kegan Paul, Trench, Trübner & Co., Ltd., 1890.
———— British Fresh-water Algæ. 2 vols. London: Williams & Norgate, 1882.
———— British Desmids. London: Williams & Norgate, 1886.
Dickie, G. Notes on Algæ from Lake Nyassa, East Africa. Jour. Lin. Soc. (Bot.), XVII, 281. London, 1879. J. R. M. S., II, 608. London, 1879.
Hassall, A. H. A History of the British Fresh-water Algæ, including Desmidieæ and Diatomaceæ. 2 vols. London, 1857.
Kützing, F. G. Species Algarum, 1849.
Migula, W. Synopsis Characearum europæarum. Leipzig, 1898.
Nägeli, C. Gattungen einzelliger Algen. 1849.

Rabenhorst, Ludovico. Flora Europæa Algarum Aquæ Dulcis et Submarinæ. Sec. 1. Algas Diatomaceas complectens. Sec. 2. Algas Phycochromaceas complectens. Sec. 3, 1-20. Algas Chlorophyllaceas. Secs. 3, 21-29. Melanophyceas et Rodophyceas complectens. Lipsiæ, 1864.

Ralfs, D. The British Desmidieæ. London, 1848.

Stokes, A. C. Analytical Keys to the Genera and Species of the Freshwater Algæ and the Desmidieæ of the U. S. Edw. F. Bigelow, Portland, Conn.

Wolle, Francis. Desmids of the United States. 2d ed. Bethlehem, Pa., 1892.

——— Fresh-water Algæ of the United States. 2 vols. Bethlehem, Pa., 1887.

Wood, Horatio C. A Contribution to the History of the Fresh-water Algæ of North America. Smith. Instn., Washington, 1872.

PROTOZOA.

Bütschli, O. Protozoa. In Bronn's Klassen und Ordnungen des Thier-Reichs. Leipzig und Heidelberg, 1880-2. 3 vols.

Claparède et Lachmann. Études sur les Infusoires. Genève, 1858-61.

Delage, Y., and E. Hérouard. Traité de Zoologie concrete. I. La cellule et les Protozoaires. Paris, 1897.

Dujardin, F. Histoire Naturelle des Infusoires. 1841.

Ehrenberg, Chr. Fr. Die Infusionsthierchen als vollkommene Organismen. 1838.

Engelmann, Th. W. Zur Naturgeschichte der Infusionsthiere. Leipzig, 1862.

Hertwig, R., und E. Lesser. Ueber Rhizopoden und denselben nahe stehende Organismen. Archiv. f. Mikroskopische Anatomie. Bd. X. Supplementheft. 1874.

Killicott, D. S. Observations on Fresh-water Infusoria. Proc. Am. Soc. Microscopy. Columbus, O., 1888.

Kent, W. Saville. A Manual of the Infusoria. 3 vols. London, 1880-81.

Lankester, E. R. Protozoa. Encyc. Brit., XIX.

Leidy, J. Fresh-water Rhizopods of North America. U. S. Geol. Sur. Washington, 1879.

Perty, M. Zur kenntnis der kleinsten Lebensformen. 1852.

Stein, F. Der Organismus der Infusionsthiere. 3 Bde. 1859-78.

——— Die Infusionsthiere auf ihre Entwickelungsgeschichte untersucht. Leipzig, 1854.

Stokes, A. C. A Preliminary Contribution toward a History of the Fresh-water Infusoria of the U. S. Jour. of the Trenton Nat. His. Soc., I, Jan., 1888.

────── Some New Infusoria. American Naturalist, Jan., 1885.

ROTIFERA.

Bourne, A. C. Rotifera, in Encyc. Britan., XXI, 1886.
Delage, Y., et E. Hérouard. Traite de Zoologie concrete. Paris, 1897.
Herrick, C. L. Notes on American Rotifera. Bull. Sci. Lab. Dennison University, 43-62. Granville, Ohio, 1885.
Hudson and Gosse. The Rotifera, or Wheel-animalcules. 2 vols. London, 1886.
Jennings, H. S. The Rotatoria of the Great Lakes. Bulletin of the Michigan Fish Commission, No. 3, 1894.

CRUSTACEA.

Baird, W. The Natural History of the British Entomostraca. London: Ray Society, 1850.
Birge, E. A. Notes on the Crustacea in Chicago Water Supply, with Remarks on the Formation of the Carapace. Chic. Med. Journal and Examiner, XIV, 584-590. Chicago, 1881.
Forbes, S. A. A Preliminary Report on the Aquatic Invertebrate Fauna of the Yellowstone National Park, Wyoming, and of the Flathead Region of Montana. Bull. U. S. Fish Com. for 1891, 209-258.
Herrick, C. L. Microscopic Entomostraca. An. Rept. of the Geology and Nat. His. Sur. Minn., 1878, 81-123.

────── A Final Report on the Crustacea of Minnesota, included in the orders Cladocera and Copepoda. From the Annual Report of Progress for 1883 of the Geol. and Nat. Hist. Survey of Minnesota.

Marsh, C. Dwight. On the Cyclopidæ and Calanidæ of Lake St. Clair. Bulletin No. 5 of the Michigan Fish Commission.

────── On the Deep-water Crustacea of Green Lake. Wis. Acad. Sci. Arts and Letters, VIII, 211-213.

BRYOZOA (POLYZOA).

Allman, G. J. The Fresh-water Polyzoa. Fol. London: Ray Society, 1856.
Braem, F. Untersuchungen über die Bryozoen des süssen Wassers. Bibl. Zool., VI, 1890.

────── Die Geschlechtliche Entwickelung von Plumatella fungosa. Zoologica, XXIII.

Davenport, C. B. Cristatella: The Origin and Development of the Individual in the Colony. Bull. Mus. Comp. Zool. Harvard College, XX. 1890.
Hyatt, A. Observations on Polyzoa. Proc. Essex Inst., IV and V, Salem, 1866–1868.
Kraepelin, K. Die deutschen Süsswasser-Bryozoa. Abh. Natur-Verein, Hamburg, X, 168.
Lankester, E. R. Polyzoa, in Encyc. Brit.
Oka, A. Observations on Fresh-water Polyzoa. Jour. Sci. College. Tokio, Japan, IV, 1891.
Stokes, A. C. The Statoblasts of our Polyzoa. The Microscope, IX, 1889.

SPONGIDÆ.

Bowerbank, J. S. Monograph of the Spongillidæ. Proc. Zoöl. Soc. London, 1863.
—— On the British Spongiadæ. 3 vols. London: Ray Soc., 1864–1874.
Carter, H. J. Note on Spongilla fragilis Leidy and on a new species of Spongilla (Mackay's) from Nova Scotia. Ann. Mag. Nat. Hist., XV, 18.
—— On a Variety of the Fresh-water Sponge Meyenia fluviatilis. *Ibidem*, 453–456, 1885.
Dawson, G. M. On some Canadian Species of Spongilla. Canadian Naturalist, N.S., VIII, 1–5, 1878.
Goetta, A. Untersuchungen zur Entwickelungsgeschichte von Spongilla fluviatilis. Hamburg, u. Leipzig, 1886.
Kellicott, D. S. The Mills Collection of Fresh-water Sponges. Bull. Buffalo Soc. Nat. Sc., V, 99–104, 1891.
Kraepelin, K. Die Fauna der Hamburger Wasserleitung. Abh. Naturw. Verein, Hamburg, XI.
Maas, O. Zur Metamorphose der Spongillalarve. Zool. Anzeiger, XII, 483–487, 1889.
Mackay, A. H. Fresh-water Sponges of Canada and Newfoundland. Proc. Trans. Roy Soc. Canada, VII, p. 85–95. 1889.
Mills, H. A New Fresh-water Sponge, Heteromeyenia radiospeculata. The Microscope, 1888, 52.
—— Notes on the Spongillidæ of Buffalo. Bull. Buffalo Soc. Nat. Sci., IV, 56–60.
Potts, E. Fresh-water Sponges of Fairmount Park. Proc. Acad. Nat. Sci., Philadelphia, 1880, 330, 331.

Potts, E. On Fresh-water Sponges. Proc. Acad. Nat. Sci., Philadelphia, 1880, 356, 357.
────── Sponges from the Neighborhood of Boston. *Ibid.*, 1882, 69, 70.
────── Our Fresh-water Sponges. American Naturalist, 1883, 1293-1296.
────── On the Wide Distribution of some American Sponges. Proc. Acad. Nat. Sci., Philadelphia, 1884, 215-217.
────── Contributions toward a Synopsis of the American Forms of Fresh-water Sponges, with Descriptions of those named by other authors and from all parts of the world. Proc. Acad. Nat. Sci. Philadelphia, 1887, p. 158-279.
────── Biology of Fresh-water Sponges. Amer. Monthly Micr. Jour., IX, 43-46, 74-77. 1888.
────── Fresh-water Sponges. The Microscope, 1890.
Weltner, W. Spongillidenstudien. Litteratur über Spongilliden. Arch. f. Naturgesch., 209-244, 245-284. 1893.

MICROSCOPIC ORGANISMS AND WATER SUPPLIES.

Allman, Geo. Jas. On Microscopic Algæ as a Cause of the Phenomenon of the Coloration of Large Masses of Water. Physiologist (England), IV, 1852.
Arthur, J. C. A Supposed Poisonous Sea-weed in the Lakes of Minnesota. Science, II, 333.
Attfield, D. Harvey. The Probable Destruction of Bacteria in Polluted River Water by Infusoria. Brit. Med. Jour., June 17, 1893.
Babcock, H. H. On the Effect of the Reversal of the Chicago River on the Hydrant Water. Lens, I, 103, 1872.
Barrows, Walter B. Plants in City Water. Rept. Bd. Water Commissioners, Middletown, Conn., 1885, 10, 11.
Beale, Lionel S. The Constituents of Sewage in the Mud of the Thames. Roy. Micr. Soc. Jour., Feb., 1884.
Bennett, A. W. On Vegetable Growth as Evidence of Purity or Impurity of Water. St. Thomas Hospital Reports, XV (xx.). Reprint, 1892.
Bentivegna, R., and Sclavo, A. Un caso d'inquinamento in una conduttura di acqua potabile per lo sviluppo della Crenothrix, Roma, 1890. Riv. d'ig. e. san. publ. Roma, 1890.
Bischoff. Bericht über die chemische und mikroskopischen Untersuchungen der Wasser der Tegeler Anlage. Berlin, 1879.
Bokorny. Purification of Streams by Chlorophyllaceous Plants. Arch. für Hyg., XX, 1894; Jour. Roy. Micr. Soc., Dec., 1894, 714.

Boston Water Works. Annual Reports. Each report contains a summary of the work of the biological laboratory, with tables of temperature, color, micro-organisms, rainfall, etc.
  1892. Temperature Curves for Lake Cochituate. Reference to the Standard Unit. Odor caused by Synura.
  1893. Reference to Standard Unit. Note on the color of the water. Description of Synura and its effect on the water. Description of a new colorimeter. An investigation of the cause of the color of natural water.
  1894. An account of stagnation phenomena in Lake Cochituate. Note on the seasonal distribution of the Diatomaceæ and Infusoria. A key to the Infusoria found in the Boston water supply. The bleaching effect of sunlight on the coloring matter of water.
  1895. The effect of light on the growth of diatoms.
Breckenfeld, A. H. An Infusorian in the Water of San Francisco. Am. Mo. Micro. Jour., V, 1884.
Brush, C. B. Deterioration of Water in Reservoirs and Conduits, its Causes and Modes of Prevention. Rept. Bd. of Health of New Jersey, Trenton, 1889-90, XIV., 107-110.
Calkins, Gary N. On Uroglena, a genus of colony-building infusoria observed in certain water supplies of Massachusetts. 23d Ann. Rept. of Mass. St. Bd. of Health.
Campbell, Douglas N. Plants of the Detroit River. Bull. Torrey Bot. Club, 1886.
Chamberlain, C. W. Organic Impurities in Drinking Water. Ann. Rept. Conn. St. Bd. of Health, 1883.
Cohn, F. Ueber den Brunnenfaden (Crenothrix polyspora), mit Bemerkungen über die Mikroskopische Analyse des Brunnenwassers. In Beiträge zur Biologie der Pflanzen, 1870.
Conn, H. W. Report on Uroglena in Middletown, Conn., in 24th Ann. Rept. of Middletown Water Commissioners for year ending Dec. 31, 1889. Also paper by Prof. Williston.
Connecticut State Board of Health Reports for 1891, 1894, *et seq.*, contain results of monthly analyses of the water supplies of the State.
  1891. Report on the Examination of Certain Connecticut Water Supplies. By S. W. Williston, H. E. Smith, Thos. G. Lee, and Chas. J. Foote.
  1894. Report on the Sanitary Condition of the New Haven Water Supply in May, 1894.
  —— Report of the Investigations of Rivers Pollution and Water Supplies. By Prof. H. E. Smith.

Connecticut State Board of Health Reports—*Continued.*
1895. Report of the Investigations of Rivers Pollution and Water Supplies. By H. E. Smith.
1896. Ditto.
——— Description of Connecticut Public Water Supplies. (Statistics.)
Currier, Chas. G. Self-Purification of Flowing Water and the Influence of Polluted Water in the Causation of Disease. With discussion. Trans. A. S. C. E., XXIV, Feb., 1891.
Cutter, E. Suspicious Organisms in Croton Water. Med. Rec., XXI, 365-368. New York, 1882.
Davis, Floyd. An Elementary Handbook of Potable Water. Boston: Silver, Burdett & Co., 1891.
De Borbo's, V. Sur la peste des eaux du lac Balaton. Bulletin de la Société Hongroise de Géographie.
Drown, T. M. The Odor and Color of Surface Waters. Jour. N. E. W. W. Assoc., March, 1888.
——— Report to Board of Health of the City of Newport, R. I., on the Condition of its Water Supply. Boston, 1892.
Duclaux. Spontaneous Purification of River Water. Ann. Inst. Pasteur, VIII., 117, 1894.
Edwards, A. Mead. Diatoms in Croton Tap Water. Quar. Jour. Micro. Sci., X, 280.
Farlow, W. G. Reports on Peculiar Condition of the Water Supplied to the City of Boston. Report of the Cochituate Water Board, 1876.
——— Reports on Matters connected with the Boston Water Supply. Bulletin of Bussey Inst., Jan., 1877.
——— Remarks on Some Algæ found in the Water Supplies of the City of Boston, 1877.
——— On Some Impurities of Drinking Water Caused by Vegetable Growths. Supplement to 1st Ann. Rept. Mass. St. Bd. of Health. Boston, 1880.
——— Some Impurities of Drinking Water. Boston: Rand, Avery & Co., 1880.
——— Relations of Certain Forms of Algæ to Disagreeable Tastes and Odors. Science, II, 333, 1883.
FitzGerald, Desmond. Spongilla in Main Pipes. Trans. A. S. C. E., XV, 337.
Forbes, F. F. A Study of Algæ Growths in Reservoirs and Ponds. Journal of the N. E. Water Works Assoc., IV, June, 1890. Reprinted in Fire and Water, July 19, 1890.
——— The Relative Taste and Odor Imparted to Water by Some Algæ and Infusoria. Jour. of the N. E. Water Works Assoc., VI, June, 1891.

Fteley, A. Algæ in a Water Supply. Supplement to 1st Ann. Rept. of Mass. St. Bd. of Health, Lunacy, and Charity, 1880.

Fuller, George W. Report on the Investigation into the Purification of the Ohio River Water at Louisville, Ky. New York: D. Van Nostrand Co., 1898.

Garrett, J. H. The Spontaneous Pollution of Reservoirs. (Odor produced by Chara.) Lancet, Jan. 7, 1893.

—— Crenothrix polyspora, var. Cheltonensis. A history of the reddening and contamination of a water supply and of the organism which caused it, with general remarks upon the coloration and pollution of water by other algæ. Public Health, IX, 15-21. London, 1896-97.

Giard, A. Sur le Crenothrix Kuhniana; la cause de l'infection des eaux de Lille. Compt. rendu Acad. d. Sc., XCV, 247-249. Paris, 1882.

Gissler, C. F. Contributions to the Fauna of the New York Croton Water. Microscopical Observations during the years 1870-71. New York, 1872.

Gray, W. Notes on the Proposal to Erect a Roof over the Malabar Hill Reservoir. Tr. M. and Phys. Soc. Bombay, 1882.

Hassall, Arthur H. The Diatomaceæ in the Water Supplied to the Inhabitants of London: Microscopic Examination of the Water. London, 1856.

Hazen, Allen. The Filtration of Public Water Supplies. New York: John Wiley & Sons, 1895.

—— Report on the Mechanical Filtration of the Public Water Supply of Lorain, Ohio. Ohio Sanitary Bulletin, I, Oct., 1897.

Hill, John W. The Purification of Public Water Supplies. D. Van Nostrand, 1898.

——Bacteria and Other Organisms in Water. Trans. Am. Soc. Civ. Eng., XXXIII, 423-466, May, 1895.

Hill, Hibbert, and J. W. Ellms. Report of the Rockville Centre Laboratory of the Department of Health of the City of Brooklyn. 1897.

Hitchcock, R. Croton Water in August. Am. Mo. Micro. Jour., II, 156, 157. New York, 1881.

Horsford, E. N., and Chas. T. Jackson. Report on the Disagreeable Tastes and Odors in the Cochituate Water Supply. Ann. Rept. Coch. Water Bd., 1854.

Hyatt, J. D. Sporadic Growth of Certain Diatoms and the Relation thereof to Impurities in the Water Supply of Cities. Proc. Am. Soc. Microscopy, 197-199. 1882.

Hueppe, F. Die hygienische Beurtheilung des Trinkwassers vom biologischen Standpunkte. Schilling's Journal für Gasbeleuchtung und Wasserversorgung. 1887.

Index Catalogue of the Library of the Surgeon-General's Office, U. S. A. Washington, 1895. Vol. XVI contains an extensive bibliography of Water and Water Supply. Continued in the Index Medicus.

Jackson, D. D., and J. W. Ellms. On Odors and Tastes of Surface Waters, with special reference to Anabæna. Technology Quarterly, X, Dec., 1897.

Jelliffe, Smith Ely. Preliminary Report on the Vegetable Organisms found in the Ridgewood Water Supply. Bulletin of the Torrey Botanical Club. New York, June, 1893.

—— A Preliminary Report upon the Microscopical Organisms found in the Brooklyn Water Supply. The Brooklyn Medical Journal, Oct., 1893.

—— A Further Contribution to the Microscopical Examination of the Brooklyn Water Supply. The Brooklyn Medical Journal, Oct., 1894.

Jelliffe, Smith Ely, and Karl M. Vogel. A Report upon Some Microscopical Organisms found in the New York City Water Supply. New York Medical Journal, May 29, 1897.

Kean, A. L., and E. O. Jordan. A Glass of Water. A brief description of the organisms in Boston tap water. Technology Quarterly, Feb., 1889.

Kellicott, S. D. Notes on Microscopic Life in the Buffalo Water Supply. Am. Jour. Micr. Pop. Science, III, 200-250. New York, 1878.

Lattimore, S. A. Report on the Recent Peculiar Condition of the Hemlock Lake Water Supply. Ann. Rept. of Ex. Bd. Rochester, N. Y., 1876-77.

Le Conte, L. J. Some Facts and Conclusions bearing upon the Relations Existing between Vegetable and Animal Growths and Offensive Tastes and Odors in Certain Water Supplies. Proc. Am. Water Works Assoc. 1891.

Leeds, Albert R. Report on the Results of the Chemical and Microscopical Examination of the Water Supply of Brooklyn, N. Y. Published by the Department of City Works, May 1, 1897.

—— Final Report of a Chemical Investigation of the Water Supply of Philadelphia. Ann. Rept. Chief Eng., 1885, 379-400.

—— Report of the Committee on Animal and Vegetable Growths affecting Water Supplies. Proc. Am. Water Works Assoc., 11th Ann. Meeting, 1889. Second Rept. in Proc. 12th Ann. Meeting.

Lemmermann, E. Die Algenflora der Filter des bremischen Wasserwerkes. Abhandl. d. naturw. Vereins zu Bremen, XIII. Band, 2 Heft, 1895.

Lewis, W. B. Report on the Microscopical Examination of the Croton and Ridgewood Waters. 4th Ann. Rept. Metropolitan Board of Health. New York, 1869.
Lewis, W. J. Microscopical Examination of Potable Waters in the State of Connecticut. Rept. Bd. Health, Conn., 1882.
Macadam, Ivison. Note on the Presence of Certain Diatoms in a Town Water Supply. Proc. Roy. Phys. Soc. Edin., 483. Edinburgh, 1885. J. R. M. S., VI, ser. 2, 291. London, 1886.
Mason, Wm. P. Water Supply. New York: John Wiley & Sons, 1896.
Mass. State Board of Health. Special Reports.
  1890. Special Report on Examination of Water Supplies.
    Examination of Water Supplies and Rivers.
    The Chemical Examination of Waters and the Interpretation of Analyses. Dr. T. M. Drown.
    Report upon the Organisms, except the Bacteria, found in the Waters of the State. G. H. Parker.
    Summary of Water-Supply Statistics—Rainfall, Flow of Streams, Temperature of Air and Water. F. P. Stearns.
    A Classification of the Drinking Waters of the State.
    Special Topics relating to the Quality of Public Water Supplies— The Effect of Storage, Investigation of Deep Ponds, Special Characteristics of Certain Surface Waters, The Natural Filtration of Water. F. P. Stearns and T. M. Drown.
    The Pollution and Self-Purification of Streams. F. P. Stearns.
  1890. Special Report on Purification of Sewage and Water.
    Filtration of Sewage and Water, and Chemical Precipitation of Sewage. Hiram F. Mills.
    A Report of the Chemical Work of the Lawrence Experiment Station. T. M. Drown and Allen Hazen.
    Experiments upon the Chemical Precipitation of Sewage at the Lawrence Experiment Station. Allen Hazen.
    A Report of the Biological Work of the Lawrence Experiment Station. Wm. T. Sedgwick.
    Investigations upon Nitrification and the Nitrifying Organism. E. O. Jordan and Ellen H. Richards.
  1895. Special Report upon a Metropolitan Water Supply for Boston.
    Improvement of the Quality of the Sudbury River Water by the Drainage of the Swamps upon the Watershed. Desmond Fitz-Gerald.
    On the Amount and Character of Organic Matter in Soils and its Bearing on the Storing of Water in Reservoirs. T. M. Drown.

Mass. State Board of Health. Annual Reports.
The annual reports since 1890 contain reports upon the examination of water supplies and experiments on the filtration of sewage and water, besides the following papers:
1890. Suggestion as to the Selection of Sources of Water Supply. By F. P. Stearns.
1891. On the Amount of Dissolved Oxygen contained in Waters of Ponds at Reservoirs at Different Depths. T. M. Drown.
The Effect of Aeration of Natural Waters. T. M. Drown.
The Microscopical Examination of Water. Gary N. Calkins.
The Differentiation of the Bacillus of Typhoid Fever. G. W. Fuller.
On Uroglena. Gary N. Calkins.
1892. Interpretation of Water Analyses. T. M. Drown.
On the Amount of Dissolved Oxygen in the Water of Ponds and Reservoirs at Different Depths in Winter, under the Ice. T. M. Drown.
On the Mineral Contents of Some Natural Waters in Massachusetts. T. M. Drown.
A Study of odors observed in the Drinking Waters of Massachusetts. Gary N. Calkins.
Seasonal Distribution of Microscopic Organisms in Surface Waters. Gary N. Calkins.
Some Physical Properties of Sands and Gravels with Special Reference to their Use in Filtration. Allen Hazen.
Reports on Epidemics of Typhoid Fever in Massachusetts in 1892. Wm. T. Sedgwick.
1893. On the Amount and Character of Organic Matter in Soils, and its . Bearing on the Storage of Water in Reservoirs. T. M. Drown.
The Filter of the Water Supply of the City of Lawrence and its Results. Hiram F. Mills.
1894. The Composition of the Water of Deep Wells. T. M. Drown.
The Bacterial Contents of Certain Ground Waters. W. T. Sedgwick.
Physical and Chemical Properties of Sand. H. W. Clark.
Report upon an Epidemic of Typhoid Fever in Marlborough. Wm. T. Sedgwick.
1895. The Hardness of Water and Methods by which it is Determined. Ellen H. Richards.
Methods Employed for the Quantitative Determination of Bacteria in Sewage and Water. G. W. Fuller and W. R. Copeland.
1896. No special papers on water or sewage analysis.
1897. No special papers on water or sewage analysis.

McElroy, Samuel. Organic Life and Matter in Water. Proc. Am. Water Works Assoc. 11th Annual Meeting, 1887.

Mills, H. Micro-Organisms in Buffalo Water Supply and in Niagara River. Proc. Am. Soc. Micrs., 1882, 165-175.

Moriez, R. L'odeur des cours d'eau au square Vauban à Lille. Rev. biol. du nord de la France, 1893-4; 55-61.

Nichols, William R. Report of the Examination of Mystic Pond and its Sources of Supply. Rept. Mass. State Board of Health, II, 1871, 387-390.

——— On the Present Condition of Certain Rivers of Massachusetts, etc. Rept. Mass. State Board of Health, V, (1874), 61-152.

——— (with W. G. Farlow and E. Burgess). On a Peculiar Condition of the Water Supplied to the City of Boston, 1875-76. Rept. Coch. Water Board, 1876.

——— Report on Matters connected with the Boston Water Supply. Rept. Boston Water Board, I, 1877, 11-15.

——— Circular of Information about Certain Fresh-water Algæ. [Printed for private distribution.] Also, in Rept. Cambridge Water Board, 1877, 8-13.

——— On the Condition of the Water of Springfield, Mass., during 1876. Rept. of Water Commissioners, Springfield, 1877.

——— Report on a Peculiar Taste and Odor of the New London (Conn.) Water. Rept. New London Water Commissioners, 1880, 27-30.

——— Tastes and Odors of Surface Waters. Jour. Assoc. of Eng. Soc., Jan., 1882.

Ohio State Board of Health. Preliminary Report of an Investigation of Rivers and Deep Ground Waters of Ohio. 1897-8.

Parker, G. H. Report on the Organisms, excepting the Bacteria, found in the Waters of the State, June, 1887—July, 1889. Mass. State Board of Health. Special Rept. on Examination of Water Supplies. Boston, 1890.

Phipson, T. L. Sur deux substances, la palmelline et la characine extraites des algues d'eau douce. Comptes Rendus, II, 1078, 1879.

Potts, E. Fresh-water Sponges as Improbable Causes of the Pollution of River Water. Proc. Ac. Nat. Sc., Phila., 1884, 30.

Rafter, G. W. On the Micro-Organisms in Hemlock Water. Rochester, 1888.

——— Some of the Minute Animals which Assist the Self-Purification of Running Streams. Discussion of a paper by Chas. G. Carrier on the Self-Purification of Flowing Water and the Influence of Polluted Water in the Causation of Disease. Trans. A. S. C. E., XXIV, Feb., 1891.

——— Some of the Circumstances affecting the Quality of a Water Supply. Proc. Am. Water Works Assoc., 12th Ann. Meeting, 1892.

Rafter, G. W. (and M. L. Mallory and J. Edw. Lane) Volvox globator as the cause of the fishy taste and odor of the Hemlock Lake Water in 1888. Ann. Rept. of Ex. Bd. of Rochester, N. Y., for 2 yrs. ending April 1, 1889.

—— On the Fresh-water Algæ and their Relation to the Purity of Public Water Supplies, with discussion. Trans. Am. Soc. of Civil Eng., Dec., 1889.

—— Deterioration of Water in Reservoirs; its Causes and Prevention. 14th Ann. Rept. N. J. St. Bd. of Health, 1890.

—— On Lake Erie as a Water Supply for the Towns on its Borders. Buffalo Medical Journal, Aug., 1896. Read before Micro. Club of Buffalo Soc. of Nat. Sciences, Jan. 13, 1896.

Reading, Pa. Report of the Board of Water Commissioners on the Purification of the Water Supply by Filtration. Report of Emil L. Nuebling. Report of Allen Hazen. Feb. 28, 1898.

Remsen, Ira. On the Impurity of the Water Supply. (Odor caused by Spongilla.) Boston, 1881.

Richards, Ellen H., and Isabel F. Hyams. The Composition of Oscillatoria prolifica (Greville) and Its Relation to the Quality of Water Supplies. Abstract in Proc. of Am. Soc. Adv. Sci., Aug., 1898.

Rochester, N. Y. Monthly Reports of the Board of Health contain results of analyses of Hemlock Lake Water Supply.

Scott, W. B., and others. Water and Water Supply. A series of papers. International Health Exhibition, 1884.

Sedgwick, Wm. T. Utilization of Surface Water for Drinking Purposes. Jour. N. E. W. W. Assoc., Sept., 1890.

—— Report on the Biological Work of the Lawrence Experiment Station. Spec. Rep. of the Mass. State Board of Health. Pt. II, 808. Boston, 1890.

—— The Data of Filtration. Pt. I. On some recent experiments on the removal of bacteria from drinking water by continuous filtration through sand. Technology Quarterly, III, 69, 1890.

—— The Data of Filtration. Pt. II. On Crenothrix Kuehniana. Technology Quarterly, III, 338, Nov., 1890.

—— The Purification of Drinking Water by Sand Filtration. Jour. N. E. W. W. Assoc., VII.

Sedgwick, Wm. T., and S. R. Bartlett. A Biological Examination of the Water Supply of Newton, Mass. Proc. Soc. Arts, M. I. T., 1887-88, 46. Also, Technology Quarterly, I, 272.

Smart, C. Report on the Water Supply of Mobile and New Orleans. Rept. Nat. Bd. of Health, 1880. 441-514.

Smith, H. E. Report on Chemical Examination of Waters, subsect. Odors and Tastes. Conn. St. Bd. of Health, 1891.

Sorby, H. C. Report of Microscopical Investigation of Sewage in Thames Water Supply. Report of Royal Commission on Metropolitan Sewage Discharge, II, 1884.

Stearns, F. P., and T. M. Drown. Discussion of Special Topics relative to the Quality of Public Water Supplies. Mass. St. Bd. of Health, 1890.

Strohmeyer, O. Die Algenflora d. Hamburger Wasserwerkes. Leipzig, 1897. Bot. Centralbl., 1898, 406.

Thresh, John C. Water and Water Supplies. London, 1896.

Torrey, J. Microscopical Examination of the Croton Water. Ann. Rept. Croton Aqueduct, 1859, 28-33.

Thomas, B. M. Microscopical Examination of the Water of Lake Michigan. Third Ann. Rept. of Dept. of Public Works. Chicago, 1879.

Traverse City, Mich. Report of the Special Water Supply Committee, with reports of engineers, H. E. Northrop, City Eng.; Geo. W. Rafter, Cons. Eng., 1897.

De Vries, Hugo. Die Pflanzen und Thiere in den dunkeln Räumen der Rotterdamer Wasserleitung. Jena, 1890.

Waller, E. Report on Croton Water during the years 1876-77. Am. Monthly Micro. Jour., II, 238. N. Y., 1881.

Weston, Robt. S. The Occurrence of Cristatella in the Storage Reservoir at Henderson, N. C. Jour. N. E. W. W. Assoc., XIII, Sept., 1898.

Whipple, Geo. C. Biology for Civil Engineers. Proc. of Society for Promotion of Engineering Education, 1896.

—— Some Observations on the Growth of Diatoms in Surface Waters. Technology Quarterly, VII, Oct., 1894.

—— Synura. Am. Monthly Micro. Jour., XV, Sept. 1894.

—— Report on Organisms in Boston Water Supply. In 19th Ann. Rept. of Boston Water Works for year ending Jan. 31, 1895.

—— Some Experiments on the Growth of Diatoms. Technology Quarterly, IX, June-September, 1896. See also Jour. N. E. W. W. Assoc., XI.

—— Raphidomonas (an organism that caused trouble in the Lynn water supply). Jour. N. E. W. W. Assoc., XI, June, 1897.

—— Biological Studies in Massachusetts. Three papers in the American Naturalist. No. 1, June 1, 1897; No. 2, July 1, 1897; No. 3, Dec., 1897.

—— Some Observations on the Growth of Organisms in Water Pipes. Jour. N. E. W. W. Assoc., Vol. XII., No. I, 1898.

Whipple, Geo. C. Interpretation of Water Analyses. A paper in the Columbia Engineer, 1897-98, 62.
White, R. Microscopic Organisms in Cochituate. Boston M. & S. J., 1878, XCIX, 4, 41.
Williston, S. W., H. E. Smith, and T. G. Lee. Report on the Examination of Certain Connecticut Water Supplies, with a Description of Certain Water Bacteria by Chas. J. Foote. 14th Ann. Rept. Conn. St. Bd. of Health. New Haven, 1892.
Zopf, W. Entwickelungsgeschichtliche Untersuchung über Crenothrix polyspora, die Ursache der Berliner Wassercalamität. Berlin, 1879.

PLANKTOLOGY.

Apstein, Carl. Das Plankton des Süsswassers und seine quantitative Bestimmung Apparate. Schriften d. naturw. Vereins f. Schleswig-Holstein, IX, 267-273.
—— Ueber die quantitative Bestimmung des Plankton im Süsswasser in Zacharias' Thier- und Pflanzenwelt des Süsswassers. 1891.
—— Quantative Plankton-Studien im Süsswasser. Biol. Centralbl. XII, 484-512.
—— Vergleich der Planktonproduktion in verschiedenen holsteinischen Seen. Bericht. d. naturf. Ges. Freiburg i. Br., VIII, 79, 80.
—— Das Süsswasserplankton, Methoden und Resultate der quantitativen Untersuchung. Kiel and Leipzig: Lipsius & Tischer, 1896.
Asper, G., and J. Heuscher. Zur Naturgeschichte der Alpenseen. Jahresber. d. St. Gallischen naturf. Ges., 1885-86 and 1887-88.
Birge, E. A. Plankton Studies on Lake Mendota. I. The vertical distribution of the pelagic crustacea during July, 1894. Trans. Wisconsin Acad. of Sci., Arts, and Letters, X, June, 1895. II. The crustacea of the plankton, July, 1894—December, 1896. Loc. cit., XI, 274-448, Dec., 1897.
—— The Vertical Distribution of the Limnetic Crustacea of Lake Mendota. Biol. Centralblatt, XVII, 371-374.
Blochmann. Die mikroskopische Thierwelt des Süsswassers. 1891.
Borne, M. von den. Das Wasser für Fischerei und Fischzucht. Neudamm, 1887.
Brandt. Ueber das Stettiner Haff. Wissenschaftl. Meeresuntersuchungen, herausgegeben von der Kommission zur Untersuchung der deutschen Meere in Kiel und d. Biolog. Anstalt auf Helgoland. Neue folge, Bd. 2, 1895.

Brandt. Ueber Anpassungserscheinungen und Art der Verbreitung von Hochseetieren. Reisebericht d. Plankton-Expedition. Ergebnisse der Plankton-Expedition, 1892.

Brooks, W. K. The Origin of the Food of Marine Animals. Bull. U. S. Fish Com., XIII, 87–92.

Brun, J. Végétations pélagiques et microscopiques du lac prés Genève. Arch. Sc. Phys. et Nat. 3 pér. XI, 543, 1884.

Cori, C. J. Ueber die Verwendung der Centrifuge in der zoologischen Technik und Beschreibung einer einfachen Handcentrifuge. Zeitschrift. f. wiss. Mikr., XIII.

Courroux, E. S. On Diatoms in the Stomachs of Shell Fish and Crustacea. Jour. Micro. and Nat. Sci., IV, 196. London, 1885. Q. J. M. S., V, 134. London, 1885.

Dean, Bashford. The Food of the Oyster; its conditions and variations. Sec. Rep. of the Oyster Investigation and of Survey of Oyster Territory for the years 1885–86. Albany, 1887, Sup. 49–78.

Dolley, C. S. The Planktonokrit, a Centrifugal Apparatus for the Volumetric Estimation of the Food Supply of Oysters and other Aquatic Animals. Proc. Acad. Nat. Sci., Philadelphia, 1896, 276-80.

Eigenmann, C. H. Turkey Lake as a Unit of Environment and the Variation of its Inhabitants. Proc. Indiana Acad. Sci., V, 204–296.

Field, Geo. W. Methods in Planktology. 10th An. Rept. Rhode Island Agricultural Experiment Station, 1897.

—— Use of the Centrifuge for Collecting Plankton. Science, N. S., VII, 163.

Forbes, S. A. The first Food of the Common Whitefish. Bull. Ill. State Lab., I, 95–109.

Fordyce, Chas. A New Plankton Pump. Proc. Nebraska Hist. Soc., 2d Series, II, 1898.

Forel, F. A. Les Micro-organismes pélagiques des lacs de la région subalpine. Bull. d. l. soc. vaud. d. sc. nat. 3 sér., XXIII, 1888.

—— Étude sur les variations de la Transparence des eaux du lac Léman. Archives d. scienc. phys. et. nat., 1877, 59.

—— Faunistische Studien in den Süsswasserseen der Schweiz. Zeitsch. f. wissensch. Zoologie, 1878. XXX, Suppl.

Francé, R. H. Zur Biology des Planktons. Vorläufige Mittheilung. Biol. Centralbl., XIV, 33–38.

Frenzel. Die biologische Fischerei-Versuchs-Station Müggelsee. Zeitschr. f. Fischerei und deren Hilfswissenschaften, 1895.

Frié and Vávra. Die Thierwelt des Unterpocernitzer und Gatterschlager Teiches. Unters. ü. d. Fauna der Gewäs. Böhmens, IV.

Fritsch, Anton. Die Stationen zur Durchforschung der Süsswasserfauna in Böhmen. Wiener landw. Zeit., 1891.

Haeckel, E. Plankton-Studien. Jenaische Zeitschrift, XXV, 232-337. Translated in Rept. U. S. Fish Com., 1889-91, 565-641.

Hensen, V. Ueber die Bestimmung des Planktons oder des im Meere treibenden Materials an Pflanzen und Thieren. V. Bericht d. Kommission zur wiss. Untersuchung d. deutschen Meere zu Kiel, XII-XIV, 1-107, 1887.

—— Methodik der Untersuchungen bei der Plankton-Expedition. Kiel u. Leipzig, 1895.

Heuscher, J. Zur Naturgeschichte der Alpenseen. Jahresbericht der St. Gallischen naturf. Ges., 1888-89.

Hoppe-Seyler. Ueber die Vertheilung absorbierten Gase im Wasser der Bodensees und ihre Beziehungen zu den in ihm lebenden Thieren und Pflanzen. In Schriften des Vereins f. Gesch. d. Bodensees u. s. Umgebung, Heft 24.

Imhof, O. E. Resultate meiner Studien über die pelagische Faune der Süsswasserbecken der Schweiz. Zeitschrift f. wissenschaftl. Zoologie, XL, 151-178, 1884.

Klebahn, H. Ueber wasserblütebildende Algen und über das Vorkommen von Gasvacuolen bei den Phycochromaceen. Forschungsber. v. Plön, IV, 189-206.

—— Allgemeiner Charakter der Pflanzenwelt der Plöner Seen. Forschungsber. v. Plön, III, 1-17.

Knudsen, Martin. De l'influence du plankton sur les quantités d'oxygène et d'acide carbonique dissous dans l'eau de Mer. Comptes Rendus. T. CXIII.

Kofoid, Chas. A. On Some Important Sources of Error in the Plankton Method. Science, N. S., VI, Dec. 3, 1897.

—— Hints on the Construction of a Tow Net. Bull. Ill. State Lab. Nat. Hist., V, Art. 1. Jour. of Applied Microscopy, I.

—— The Fresh-Water Biological Stations of America. The American Naturalist, XXXII. June, 1898.

—— Plankton Studies. Methods and Apparatus. Bull. Ill. State Lab. Nat. Hist., V, 1-25, 1897.

Lankester, Ed. Ray. The Cause of the Green Color of the European Oyster. Amer. Naturalist, XX, 298. 8vo. Philadelphia, 1886.

Lauterborn. Ueber Periodicität im Auftreten und in der Fortpflanzung einiger pelagischer Organismen des Rheines und seiner Altwässer. Verh. d. nat. med. Vereins zu Heidelberg, N. F., V, 1893.

Lemmermann, E. Zur Algenflora des Plöner Seegebietes. 2. Beitrag. Forschungsber. v. Plön, III, 88-188, 1895.

Marsh, C. Dwight. On the Limnetic Crustacea of Green Lake. Trans. Wis. Acad. of Sci., Arts, and Letters, XI, Aug., 1897.

Peck, James I. On the Food of the Menhaden. Bull. U. S. Fish Com., XIII, 1893.

——— The Sources of Marine Food. Bull. U. S. Fish Com., XV, 1895.

Peck, James I., and N. R. Harrington. Observations on the Plankton of Puget Sound. Trans. N. Y. Acad. Sci., XVI. Feb., 1898.

Reighard, J. E. A Biological Examination of Lake St. Clair. Bulletin No. 4, Michigan Fish Com., 1894.

Schröter, C. Die Schwebeflora unserer Seen (Das Phytoplankton). Neujahrsblatt herausgegeben von der Naturforschenden Gesellschaft auf das Jahr 1897, XCIX.

Schröter, C., und O. Kirchner. Der Bodensee-Forschungen, neunter Abschnitt. Die Vegetation des Bodensees. Kommissions-verlag der Schriften des Vereins für Geschichte des Bodensees und seiner Umbegung von Joh. Thom. Stettner. Lindau i. B., 1896.

Schütt. Das Pflanzenleben der Hochseen. Ergebnesse der Plankton-Expedition, 1892.

——— Analytische Plankton-Studien, 1892.

Seligo, A. Hydrobiologische Untersuchungen, I. Schriften d. naturf. Ges. Danzig, N.F., VII, 43–89.

Strodtmann, S. Die Anpassungen der Cyanophyceen an das pelagische Leben. Archiv. für Entwicklungsmechanik der Organismen, I, 1895.

——— Bemerkungen über die Lebensverhältnisse des Süsswasserplanktons. Forschungsber. v. Plön, III, 145–179.

——— Planktonuntersuchungen in holsteinischen und mecklenburgischen Seen. Forschungsber. v. Plön, III, 273–287.

Susta, J. Die Ernährung des Karpfen und seiner Teichgenossen. Stettin. 252.

Tanner, Z. L. On the Appliances for Collecting Pelagic Organisms, with special reference to those employed by the U. S. Fish Commission. Bull. U. S. Fish Com., XIV, 1894.

Ward, Henry B. Continued Biological Observations. Proc. of Nebraska Historical Soc., 2d Series, II, 1898.

——— A Biological Examination of Lake Michigan. Bull. No. 6, Michigan Fish Com., 1896.

Weltner. Forschungsberichte aus der Biologischen Station zu Plön. Zeitschr. f. Fischerei u. deren Hilfswissensch., V, 1894.

Wesenburg-Lund, C. Biologiske Undersoegelser over Ferskvandsorganismer. Vid. med. natur. For., 105–168. Kjöbenhavn.

Whipple, Geo. C. A Simple Apparatus for Collecting Samples of Water at Various Depths. Science, Dec. 20, 1895.

Whipple, Geo. C. An Apparatus for Collecting Samples of Water. The Engineering Record, July 19, 1897.

Zacharias, Otto. Forschungsberichte aus der biolog. Station zu Plön. Biologische Mitteilungen, I, 1893, 27-41. Beobachtungen am Plankton des grossen Plönersees, II, 1894, 91-137. Ueber die wechselnde Quantität des Planktons im grossen Plönersee, III, 1895, 97-117. Ueber die horizontale und verticale Verbreitung limnetischer Organismen, III, 1895, 148-128. Quantitative Untersuchungen über das Limnoplankton, IV, 1896, 1-64. Ergebnisse einer biolog. Excursion an die Hochseen des Riesengebirges, IV, 1896, 65-87.

———— Fauna des grossen Plöner Sees. Forschungsber. d. biol. Station zu Plön, II, 57-64, 1894.

APPENDIX D.

GLOSSARY TO PART II.

Adoral, relating to the mouth.
Aeruginous, of the color of verdigris; blue-green.
Alate, winged.
Amylaceous, resembling starch.
Anal, relating to the anus.
Annulate, marked with rings.
Antheridia, reproductive organs supposed to be analogous to anthers.
Arcuate, bent like a knee.
Articulate, composed of joints.
Bacillar, rod-like.
Bifid, two-cleft.
Birotulate, with two recurved rounded ends.
Botryoid, clustered like a bunch of grapes.
Buccal, relating to the cheek.
Campanulate, bell-shaped.
Capitate, collected in a head.
Carapace, a hard shell.
Carinate, like a keel.
Caudal, relating to the tail.
Cervical, relating to the neck.
Chitinous, horny.
Ciliated, provided with cilia, or hair-like appendages.
Circinate, curled round, coiled, or spirally rolled up.
Cirrose, curled as a tendril.
Clathrate, perforated or latticed like a window.
Coccus, a minute spherical form.
Coenobium, a community of a definite number of individuals united in one body.
Concatenate, linked like a chain.
Connate, united congenitally.
Convolute, rolled together.
Cortical, relating to the external layers.
Crenate, notched or scalloped.
Cuneate, wedge-shaped.
Cymbiform, boat-shaped.
Cyst, a membranous sac without opening.
Dentate, toothed.
Denticulate, finely toothed.
Dichotomous, dividing by pairs from top to bottom.
Dioecious, the males and females represented in separate individuals.
Ectoderm, the external of two germinal cellular layers.
Emarginate, with a notch cut out of the margin at the end.
Encuirassed, with an indurated dorsal shield.
Encysted, enclosed in a cryst or bladder.

GLOSSARY TO PART II.

Endochrome, the coloring matter of cells.
Endoplast, the nucleus of a protozoan cell.
Fasciculate, in bundles from a common point.
Filiform, long, slender, thread-like.
Flagellate, provided with flagella, or lash-like appendages.
Foliaceous, resembling a leaf.
Forcipitate, like forceps.
Funicular, like a cord or thread.
Furcate, forked or divergently branched.
Fusiform, tapering like a spindle.
Gibbous, swollen, convex.
Gonidia, propagative bodies of small size not produced by act of fertilization.
Heterocyst, interspersed cells of a special character differing from their neighbors.
Holophytic, like a plant.
Hormogons, special reproductive bodies composed of short chains of cells, parts of internal filaments.
Hyaline, transparent.
Hyphae, filaments of the vegetative portion of a fungus.
Indurated, hardened.
Intercalated, interspersed, placed between others.
Involute, rolled inward.
Lamellated, lamellose, in layers.
Lanceolate, lance-shaped, tapering at each end.
Lenticular, like a lens.
Lophophore, an organ bearing tentacles, found on the Bryozoa.
Lorica, a hard protective coat.
Lunate, crescent-shaped.
Macrogonidia, large gonidia.
Macrospores, large spores.
Matrix, the birth cavity.
Microgonidia, small gonidia.
Monaxonic, with but one axis.
Moniliform, like a necklace, contracted at regular intervals.
Monoecious, male and female represented in one individual.
Mucronate, having a small tip.
Mycelium, the vegetative portion of a fungus.
Naviculoid, boat-shaped.
Oosphere, an ovarian sac.
Oospore, spore produced in an ovarian sac.
Oral, relating to the mouth.
Parietal, growing near the wall.
Peristome, the oral region.
Pinnatifid, shaped like a feather.
Polythecium, an assemblage of many loricæ.
Punctate, studded with points or dots.
Pyriform, pear-shaped.
Reniform, kidney-shaped.
Replicate, folded back.
Reticulate, latticed.
Retractile, capable of being drawn back.
Saccate, like a bag.
Sarcode, the primary vital matter of animal cells (Protoplasm).
Scalariform, ladder-like.
Segregate, set apart from others.
Septate, separated by partitions.
Setiform, in the form of a bristle.
Sigmoidal, S-shaped.
Sinuate, with notches or depressions.
Spermatozoids, thread-like bodies, motile, and possessing fecundative power.
Sporangium, sporange, a spore-case.
Sporocarp, the covering or capsule enclosing a spore.
Sporoderm, the covering of a spore.
Statoblasts, the winter eggs, or reproductive bodies of the Bryozoa and Spongidæ.
Striate, covered with striæ.

Styligerous, bearing styles or prominences.
Sub-, a prefix indicating "almost," or "nearly."
Suborbicular, almost spherical.
Thallus, a leaf-like expansion.
Trichocyst, a rod-like body developed in the cortical layer of some protozoa.
Trichome, the thread or filament of filamentous algæ.

Turbinate, shaped like a top.
Utriculate, inflated.
Vacuolated, containing drops or vacuoles.
Vesiculiform, bladder-like.
Zoodendrum, a bill-like colony-stalk.
Zoogonidia, gonidia endowed with motion.
Zoospores, locomotive spores.
Zygospore, a spore resulting from conjugation.

INDEX.

Absorption of light by water, 75
Acarina, 251
Achlya, 208
Acineta, 231
Acinetaria, 230
Actinophrys, 215
Algæ, 188
Amorphous matter, 29
Amœba, 213
Amphipoda, 240
Amphora, 168
Anabæna, 184
——, decomposition of, 127
——, in Cedar Swamp, 131
——, oil isolated, 124
Anacharis, 252
Anguillula, 251
Anthophysa, 217
Anuræa, 239
Anuræadæ, 239
Aphanizomenon, 185
——, growth in winter, 101
Aphanocapsa, 181
Aphanothece, 182
Arcella, 213
Aromatic odors, 124, 126
Arthrodesmus, 198
Ascomycetes, 206
Asellus aquaticus, 240
Aspergillus, 207
Asplanchna, 237
Asplanchnadæ, 237

Asterionella, 172
——, growth of, in Brooklyn, 142
——, growth of, in ground water, 141
——, weight of, 129
Attachment to filter funnels, Sedgwick-Rafter Method, 19, 20
Autumnal overturning (autumnal circulation), 61
Bacteria, in distribution pipes, 148
Bacteriological examination of water, 8
Batrachospermum, 251
Bdelloida, 236
Beggiatoa, 177
Blank, for recording results of microscopical examination, 32
Bleaching of water by sunlight, 70
Bosmina, 243
Boston Water Works Laboratory, 2
Brooklyn Water Works Laboratory, 2
Brooklyn water-supply, growth of Asterionella, 142
——, growth of Paludicella, 155
Brachionus, 239
Branchipus, 244
Botryococcus, 191
Brachionidæ, 239
Bryozoa, 245
——, growth of, in water-pipes, 152
Bursaria, 228
Canals, organisms found in, 48

Canthocamptus, 242
Cell, errors in the, 28
——, used in Sedgwick-Rafter Method, 21
Centrifuge, use of, in concentrating microscopic organisms, 6, 38
Ceratium, 223
Ceratophyllum, 252
Cercomonadina, 217
Cercomonas, 217
Chætonotus, 251
Chætophora, 203
Chætophoraceæ, 202
Chara, 203
——, odor caused by, 119
Characeæ, 203
Chemical analysis of water, 8
——, relation to the growth of microscopic organisms, 87-92
Chestnut Hill Reservoir, temperature of, 57
Chlamydomonadina, 221
Chlamydomonas, 221
Chlorine, effect of, on the growth of microscopic organisms, 88
Chlorophyceæ, 188
——, seasonal distribution of, 100
Choano-flagellata, 222
Chroöcoccaceæ, 180
Chroöcoccus, 180
Chydorus, 244
Chrysomonadina, 219
Ciliata, 224
Circulation periods in lakes, 67
Cladocera, 243
Cladophora, 201
Cladothrix, 177
Classification of diatoms, 167
—— of lakes according to temperature, 63-68
—— of Massachusetts Ponds and Reservoirs according to the microscopic organisms present, 82
—— of microscopic organisms, 156

Classification of microscopic organisms according to their abundance, 77
Classifications of miscroscopic organisms, schedule of, 33
Clathrocystis, 181
Cleaning glassware, directions for, 258
Clean watershed, definition of, 130
Closterium, 197
Cocconeis, 171
Cocconema, 169
Cocconideæ, 171
Codonella, 227
Cœlastrum, 193
Cœlomonadina, 218
Cœlomonas, 218
Cœlosphærium, 182
Cold Spring Brook, color of, 69
Coleps, 229
Collection of samples, apparatus for, 253
Color of water, 68-72
——, effect on the growth of microscopic organisms, 88
Color readings, table for transforming, 258
Color standards, 69
Colpidium, 230
Coluridæ, 238
Colurus, 239
Compressibility of water, 50
Concentrating attachment to filter-funnels, Sedgwick-Rafter Method, 19-20
Concentration of organisms by the Sedgwick-Rafter Method, 18
Conductivity, thermal, of water, 51
Conferva, 201
Confervaceæ, 201
Confervoideæ, 201
Conjugate, 196
Connecticut St. Bd. of Health, examination of water-supplies, 2
Conochilus, 235

INDEX.

Copepoda, 241
Corethra, 251
Coscinodisceæ, 175
Cosmarium, 197
Counting-cell, Sedgwick-Rafter Method, 21
Counting, methods of, by Sedgwick-Rafter Method, 30
Crenothrix, 177
——, growth in distribution-pipes, 151
——, growth of, in ground-waters, 43, 144
Cristatella, 247
——, odor caused by, 119
Crustacea, 239
——, seasonal distribution of, 103
Cryptomonas, 221
Crypto-Raphidieæ, 174
Crystal Lake, temperature of, 61
Cucumber taste, 119
Cyanophyceæ, 179
——, seasonal distribution of, 100
Cyclops, 242
Cyclotella, 175
Cylindrospermum, 184
Cymbella, 168
Cymbelleæ, 168
Cypris, 243
Cystiphoræ, 180
Cystoflagellata, 223
Daphnia, 243
Decantation error, 28
Decapoda, 240
Decomposition, odors of, 116
—— of organisms, 127
Deep ponds, organisms in, 86
Degree of concentration, 24
Density of water, 51
Desmidieæ, 196
Desmidium, 199
Diaphanometers, 72
Diaptomus, 242
Diathermancy of water, 52
Diatoma, 173

Diatomaceæ, 157
——, cell of, 157
——, cell-contents, 161
——, classification of, 167
——, external secretions, 162
——, growth of, at different depths, 97
——, markings, 160
——, movement, 163
——, multiplication of, 164
——, reproduction of, 166
——, seasonal distribution of, 94
——, shape and size, 159
——, structure of valve, 161
——, succession of, 95
Dictyosphærium, 191
Difflugia, 214
Diglena, 238
Dimorphococcus, 191
Dinobryon, 221
Dino-flagellata, 222
Disc, use of, for comparing the turbidities of water, 75
Disintegration, errors of, 27
Dissolved oxygen in Lake Cochituate, 137
Distribution-pipes, diminution of microscopic organisms in, 147-150
——, growths of organisms in, 150
Docidium, 197
Dolley, Dr. C. S., method of concentrating microscopic organisms, 6
Draparnaldia, 202
Enchelys, 229
Encyonema, 169
Endoparasites of man found in water, 11
Entomostraca, 240
Enumeration of organisms by the Sedgwick-Rafter Method, 23
Epistylis, 226
Epithemia, 171
Errors in the Sedgwick-Rafter Method, 25

Euastrum, 198
Eudorina, 195
Euglena, 218
Euglenina, 218
Euglenoidea, 217
Euglypha, 214
Eunotia, 171
Euplotes, 225
Facultative limnetic organisms, 105
Filter used in Sedgwick-Rafter Method, 15
Filter-basins, (infiltration basins,) growth of organisms in, 45
Filter-beds, growth of organisms on, 143
Filter-galleries, (infiltration galleries,) growth of organisms in, 45
Fishy odors, 124, 126
Flagellata, 216
Flosculariadæ, 235
Floscularia, 235
Forbes, F. F., method of microscopical examination, 3
Forel, Dr. F. A., studies of Lake Geneva, 5
Fragilaria, 173
Fredericella, 246
——, growth of, in water-pipes of Boston. 153
Fungi, 205
——, seasonal distribution of, 101
Funnel errors, 25
Gammarus pulex, 240
Genevan Commission, experiments on the transparency of water, 73
Geographical distribution of microscopic organisms, 77
Glenodinium, 223
Glœocapsa, 181
Glœocystis, 190
Glœothece, 182
Gomphonema, 170
Gomphonemeæ, 170
Gonium, 195
Gonyostomum, 218
Gordius, 251

Grassy odors, 124, 126
Ground-water, character of, 42
——, organisms in, 43, 44, 45
——, storage of, 141
Gymnodinium, 223
Halteria, 225
Hardness, effect of, on the growth of microscopic organisms, 91
Hassall's method of microscopical examination, 1
Hazen, Allen, method of comparing turbidities of water, 72
Heliotropism of diatoms, 98
Heliozoa, 215
Hensen's method of collecting microscopic organisms, 5
Heteromonadina, 217
Heterophrys, 215
Heterotricha, 226
Himantidium, 172
Hippuris, 252
Holotricha, 228
Horizontal distribution of microscopic organisms, 106
Hyalotheca, 199
Hydatina, 238
Hydatinadæ, 237
Hydra, 251
Hydrodictyon, 192
——, odor caused by, 119
Hypotricha, 224
Illoricata, 236
Individual Counting System, 30, 31
Infusoria, 224
International Limnological Commission, 6
Isomastigoda, 219
Isopoda, 240
Jackson, D. D., attachment to filter-funnels, Sedgwick-Rafter Method, 19
——, isolation of oil of Anabæna, 124
——, analysis of gases of decomposition of Anabæna, 127
Kean, A. L., method of microscopical examination, 3

Lake Cochituate, temperature of, 56-63
Lake Winnepesaukee, temperature of, 61
Lakes, classification of, 63
Lemna, 252
Leptomitus, 208
Leptothrix, 176
Light, effect of light on the growth of diatoms, 97
——, transmission of, by water, 68
Limnetic organisms, 105
Limnological Commission of Switzerland, 6
Limnology, definition of, 50
Littoral organisms, 105
Lobosa, 213
Loricata, 238
Lyngbya, 185
Lynn water-supply, growth of Raphidomonas, 48
Lynn Water Works Laboratory, 2
Macdonald, J. D., method of microscopical examination, 1
Macrobiotus, 251
Malacostraca, 240
Mallomonas, 221
——, peculiar case of vertical distribution, 110
Massachusetts ponds and reservoirs, microscopic organisms in, 82
Massachusetts State Board of Health, examination of water-supplies, 2
Mastigophora, 215
Mastigocerca, 238
Melicertadæ, 235
Melosira, 174
——, odor caused by, 119
Melosireæ, 174
Meridion, 173
——, odor caused by, 118
Merismopedia, 182
Meyenia, 250

Micrasterias, 198
Microcodon, 236
Microcodidæ, 236
Microcoleus, 186
Microns, table for transforming, 257
Microscopical examination of water, 8
——, as indicating sewage contamination, 10
——, as explaining the chemical analysis, 12
——, as explaining the cause of turbidity and odor of water, 13
——, as a method of studying the food of fishes, 13
Microscopical examinations, number made in New England and New York, 3
Micrometer used in Sedgwick-Rafter Method, 22
Micro-organisms, effect of, upon health, 128
——, use of the term, 9
——, relative number at various depths, 111, 112
Microscope, outfit necessary for water analysis, 22
Microcystis, 181
Monadina, 217
Monas, 217
Mt. Prospect Laboratory, 2
Mucor, 207
Myriophyllum, 251
Nais, 251
Nassula, 229
Natural odors of organisms, 117
Navicula, 169
Naviculeæ, 169
Nauplius, 242
Nematogenæ, 182
Nephrocytium, 191
Nitella, 203
Nitrogen, effect of, on the growth of microscopic organisms, 91

Nitzschia, 174
Nostoc, 183
Nostocaceæ, 183
Noteus, 239
Notholca, 239
Notommatadæ, 238
Ocular micrometer used in Sedgwick-Rafter Method, 22
Odor-producing substances in microscopic organisms, 121
Odors caused by littoral organisms, 118
—— caused by microscopic organisms, 116
—— caused by organic matter, 114
—— caused by the coloring matter of water, 115
——, chemical, 127
——, classification of odors due to organisms, 125
—— in water-supplies, 113, 128
——, methods of observing, 115
——, natural, of organisms, 117
—— of growth, 117
—— of decomposition, 116
—— of disintegration, 117
——, terms describing their intensities, 114
Œdogoniaceæ, 202
Oil of Anabæna, 124
—— of Uroglena, 124
——, the cause of odors in organisms, 122
Oils, dilution at which their odor ceases to be recognized, 123
Ophiocytium, 192
Organic matter, removal from reservoir sites, 139
Oscillaria, 177, 185
Oscillarieæ, 185
Ostracoda, 243
Oxygen, dissolved in water, 137
Palmella, 190
Palmellaceæ, 190
Paludicella, 247

Paludicella, growth of, in Brooklyn water-pipes, 155
Pandorina, 195
Paramæcium, 228
Parker, G. H., method of microscopical examination, 3
Peck, Prof. James I., studies of fish-food, 6, 13
Pectinatella, 247
——, odor caused by, 119
Pediastrum, 193
Pelagic organisms, 105
Penicillium, 206
Penium, 197
Peridinium, 223
Peritricha, 225
Phacus, 219
Phæophyceæ, 188
Philodinadæ, 236
Phycochromophyceæ, 179
Phycocyanine, 179
Phycomycetes, 207
Phyllopoda, 244
Physical examination of water, 8
Physical properties of water, 50
Phytoglœa, 29
Phytozoa, 210
Pinnularia, 170
Pipe-moss, 152
——, effect of, on capacity of water pipes, 154
Plankton, definition of, 5
Plankton Net Method, 34
Plankton pump, 7, 38
Plankton studies in America, 6
Planktonokrit, 6, 38
Pleuronema, 230
Pleurosigma, 170
Ploima, 236
Plön Biological Laboratory, 6
Plumatella, 246
Polyarthra, 237
Polyedrium, 192
Polyzoa, 245
Ponds, growth of organisms in, 132

Potamogeton, 252
Precision of the Sedgwick-Rafter Method, 28
Protococcaceæ, 191
Protococcoideæ, 190
Protococcus, 192
Protozoa, 209
———, seasonal distribution of, 101
Protozoan cell, 210
Pseudo-raphidieæ, 171
Rafter, Geo. W., improvements in the method of microscopical examination, 4
Rain-water, organisms in, 41
Raphidieæ, 168
Raphidium, 191
Raphidomonas, 218
——— in the Lynn water-supply, 48
Rattulidæ, 238
Reproduction of diatoms, 166
Reticularia, 214
Rhizopoda, 212
Rhizota, 235
Rivularia, 187
Rivularieæ, 186
River-water, organisms in, 45
Rodophyceæ, 188
Rotifer, 236
Rotifera (Rotatoria), 232
———, seasonal distribution of, 103
Saccharomyces, 206
Sampling, errors of, 25
Sand error, 26
Sand used in Sedgwick-Rafter Method, 16
Sanitary water-examination, data obtained by, 8
Saprolegnia, 207
Sarcoda, 212
Scenedesmus, 192
Schedules of classification of microscopic organisms, 33
Schizonema, 170
Schizomycetes, 176
———, seasonal distribution of, 101

Schizophyceæ, 176
Scirtopoda, 239
Scytonema, 186
Scytonemeæ, 186
Seasonal distribution of Chlorophyceæ, 100
——— of Cyanophyceæ, 100
——— of microscopic organisms, 93
Secchi, experiments on the transparency of water, 73
Sedgwick, Wm. T., improvements in the method of microscopical examination, 4
Sedgwick-Rafter method of microscopical examination, 4, 15
Seeding of reservoirs by organisms from swamps, 132
Sewage, microscopical examination of, 11
Shallow ponds, organisms in, 86
Sida, 244
Siphoneæ, 200
Sirosiphon, 186
Sirosiphoneæ, 186
Smith, H. L., classification of diatoms, 168
Sorastrum, 193
Sorby, H. C., method of microscopical examination, 2
Sphagnum, 251
Sphærozosma, 199
Sphærozyga, 184
Spirogyra, 200
Sponge, growth of, in water-pipes, 151, 152
Spongidæ, 248
Spongilla, 249
———, odor caused by, 119
Spring overturning (spring circulation), 59
Stagnant pools, growth of organisms in, 133
Stagnation, 58-60
———, effects of, 99, 134
Statoblasts, 246

Standard Unit, 29, 31
Staurastrum, 198
Staurogenia, 193
Stauroneis, 170
Stentor, 227
Stephanodiscus, 175
Stigeoclonium, 202
Storage of ground-water, 141
Storage reservoirs, low-level gate, 140
——, soil to be removed from site of, 139, 140
Storage of surface-water, 130
Stratification of water, 51
Suctoria, 230
Surface-water, organisms in, 45
——, storage of, 130
Surirella, 174
Surirelleæ, 174
Swamps, effect of, on water, 131
——, growth of organisms in, 131
Synchæta, 237
Synchætadæ, 237
Syncrypta, 220
Synedra, 172
Synura, 219
——, odor caused by, 120
Tabellaria, 173
Tabellarieæ, 173
Taste, its relation to odor, 113
Temperature of lakes and ponds, 52
—— of water in distribution pipes, 146
—— of water, methods of observation, 52
Tentaculifera, 230
Tetmemorus, 198
Tetrapedia, 182
Tetraspora, 190
Thermocline, 62
——, its relation to the vertical distribution of microscopic organisms, 110

Thermophone, 53-56
Tintinnidium, 227
Tintinnus, 226
Trachelocerca, 230
Trachelomonas, 219
Transparency of water, 73
Triarthra, 237
Triarthradæ, 237
Trinema, 214
Tubifex, 251
Turbidity of water, 72
Ulothricheæ, 202
Ulothrix, 202
Unit, Standard, 29, 31
Uroglena, 220
—— oil isolated, 124
Utricularia, 252
Uvella, 220
Vaucheria, 201
Vaucheriaceæ, 200
Vertical distribution of microscopic organisms, 107
Volvocina, 222
Volvocineæ, 193
Volvox, 194
Vorticella, 225
Water analyses, value of, 9
Water-pipes, growth of organisms in, 146
Weights and measures, conversion tables, 256
Wind, effect of, on horizontal distribution of microscopic organisms, 106
——, effect of, on vertical distribution of microscopic organisms, 109
Xanthidium, 198
Zoogloea, 29
Zoothamnium, 226
Zygnema, 200
Zygnemaceæ, 199
Zygogonium, 200

PLATE I.
DIATOMACEÆ.

PLATE I.

DIATOMACEÆ.

Magnification 500 diameters.

Fig. A. Navicula viridis, valve view.
" B. Navicula viridis, girdle view.
" C. Navicula viridis, transverse section.
    a, Outer, or older valve. b, Inner, or younger valve. c, c', Connective bands, or girdles. d, Central nodule. cc, Terminal nodules. f, Raphé. g, Furrows. m, Chromatophore plates. n, Nucleus. o, Oil globules. p, Cavities. u, Protoplasm.

Figs. D, E, F. Navicula viridis, sectional views showing multiplication by division. After Deby.
    a, Valve. b, Girdle. c, Protoplasm. d, Chromatophore plates. e, Central cavities. f, Nucleus and nucleolus. g, Oil globules.

Fig. 1. Amphora, valve view.
" 2. Amphora, girdle view.
" 3. Cymbella, valve view.
" 4. Cymbella, valve view.
" 5. Encyonema. A, valve view. B, girdle view.
" 6. Cocconema. A, valve view. B, girdle view.
" 7. Navicula gracilis, valve view.
" 8. Navicula Rhyncocephara, valve view.
" 9. Stauroneis, valve view.
" 10. Stauroneis, girdle view.
" 11. Pleurosigma, valve view.
" 12. Gomphonema. A, valve view. B, girdle view.
" 13. Cocconeis, valve view.
" 14. Cocconeis, girdle view.
" 15. Epithemia, valve view.
" 16. Epithemia, girdle view.
" 17. Eunotia, valve view.

PLATE I.

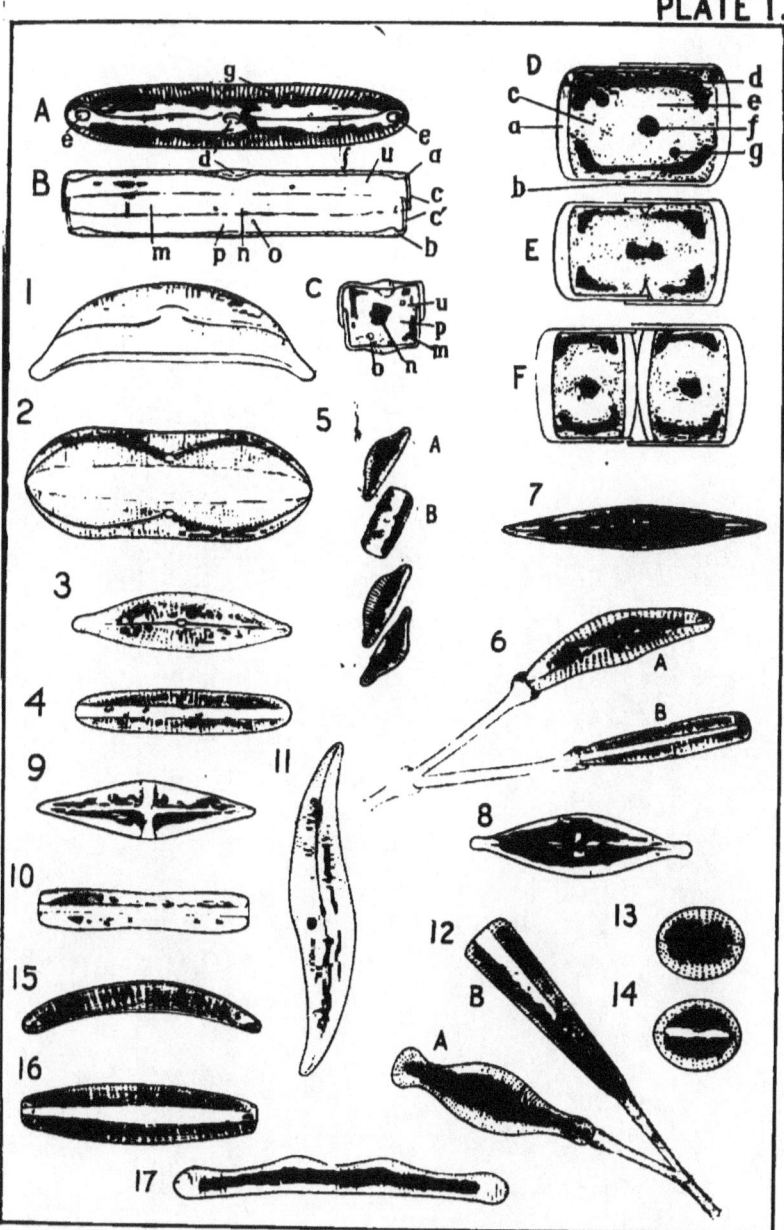

G.C.W. del.

PLATE II.
DIATOMACEÆ.

PLATE II.
DIATOMACEÆ.

Magnification 500 diameters.

Fig. 1. Himantidium, valve view.
" 2. Himantidium, girdle view.
" 3. Asterionella, valve view.
" 4. Asterionella, girdle view (typical form).
" 5. Asterionella, girdle view, showing division of the cells.
" 6. Asterionella, girdle view, showing rapid multiplication.
" 7. Asterionella. *A*, valve view. *B*, girdle view.
" 8. Synedra pulchella, valve view.
" 9. Synedra pulchella, girdle view.
" 10. Synedra ulna, valve view.
" 11. Synedra ulna, girdle view.
" 12. Fragilaria, girdle view.
" 13. Fragilaria, valve view.

PLATE II.

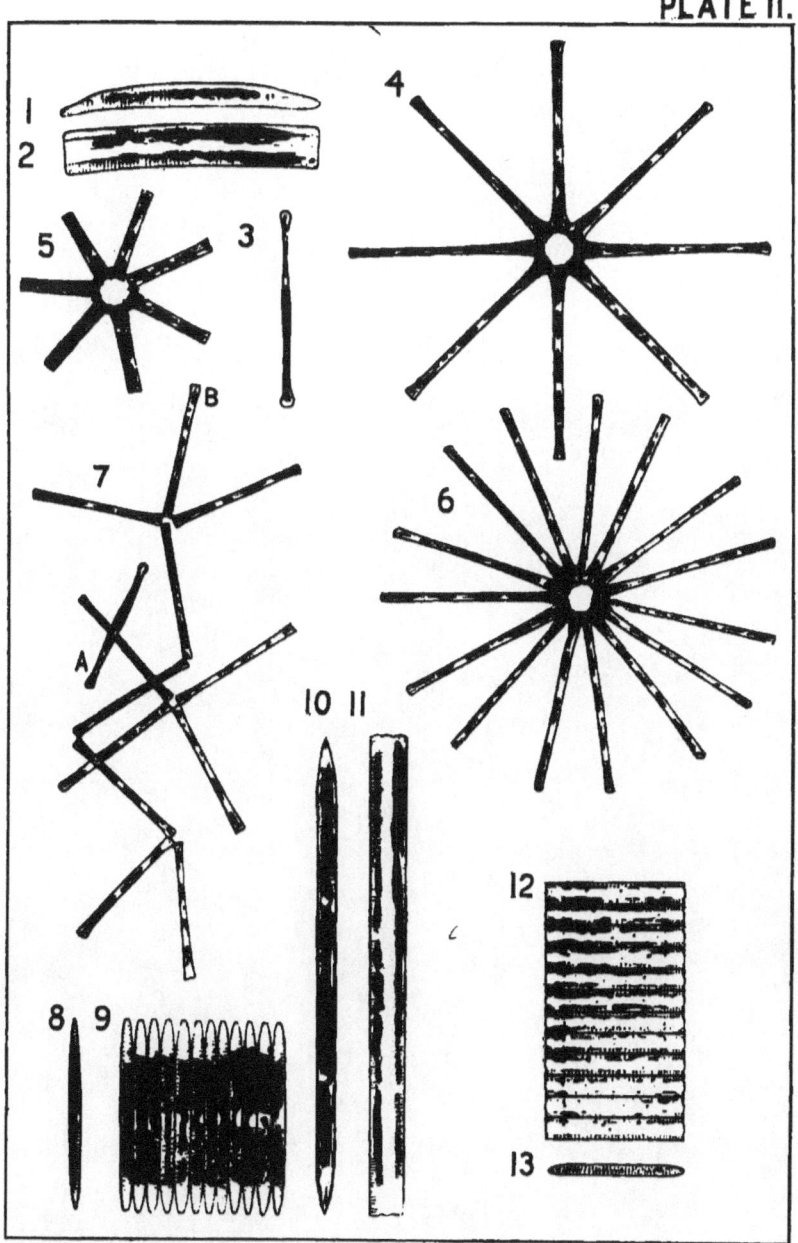

G.C.W. del.

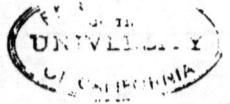

PLATE III.
DIATOMACEÆ.

PLATE III.

DIATOMACEÆ.

Magnification 500 diameters.

Fig. 1. Diatoma vulgare, valve view.
" 2. Diatoma vulgare, girdle view.
" 3. Diatoma tenue, girdle view.
" 4. Meridion circulare, valve view.
" 5. Meridion circulare, girdle view.
" 6. Tabellaria fenestrata, valve view.
" 7. Tabellaria fenestrata, girdle view.
" 8. Tabellaria flocculosa, valve view.
" 9. Tabellaria flocculosa, girdle view.
" 10. Nitzschia sigmoida, valve view.
" 11. Nitzschia sigmoida, girdle view.
" 12. Nitzschia longissima, girdle view.
" 13. Surirella, valve view.
" 14. Surirella, girdle view.
" 15. Melosira, valve view.
" 16. Melosira, girdle view.
" 17. Melosira auxospore.
" 18. Cyclotella, valve view.
" 19. Cyclotella, girdle view.
" 20. Stephanodiscus, valve view.
" 21. Stephanodiscus, girdle view.

PLATE III.

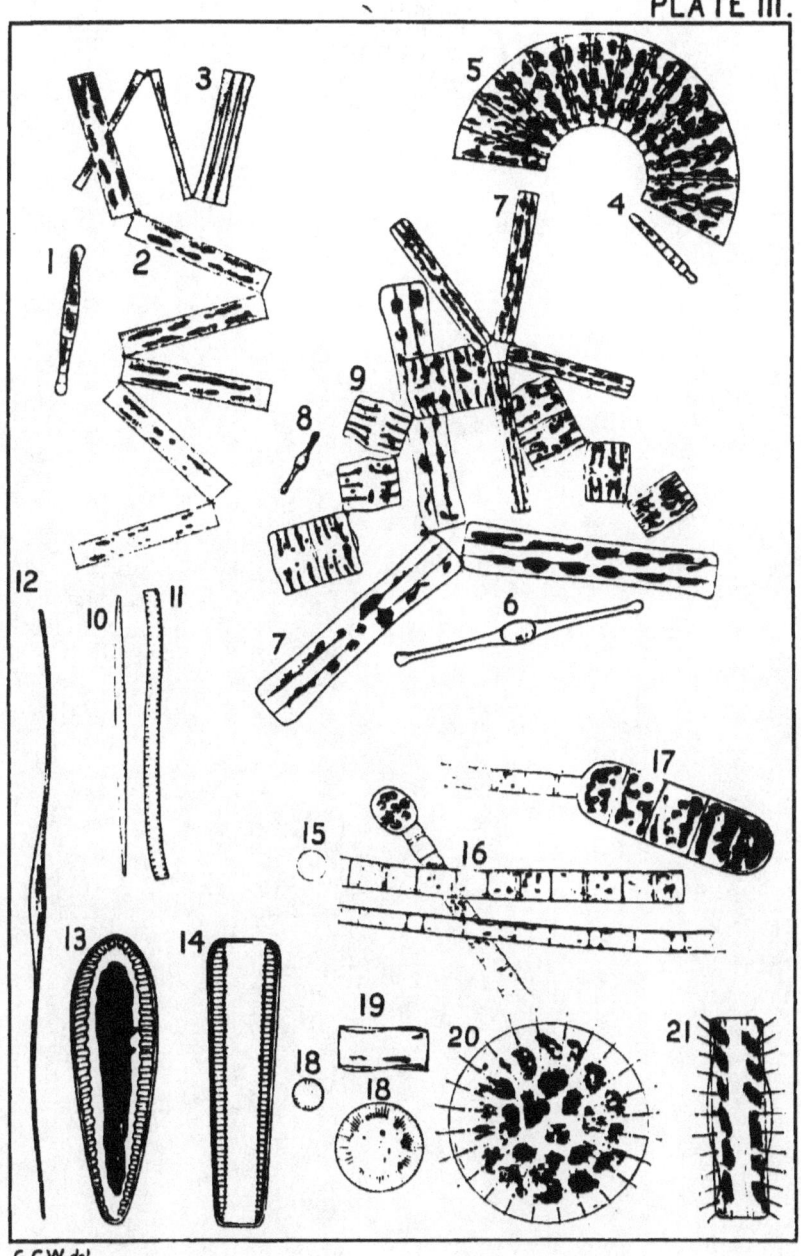

G.C.W. del.

PLATE IV.

SCHIZOMYCETES. CYANOPHYCEÆ.

PLATE IV.

SCHIZOMYCETES.

Magnification 500 diameters.

Fig. 1. Leptothrix.
" 2. Cladothrix, showing false branching.
" 3. Beggiatoa.
" 4. Crenothrix. *A*, filament enclosed in sheath. *B*, filament with sheath removed, showing liberation of spores.

CYANOPHYCEÆ.

Magnification 500 diameters.

Fig. 5. Chroöcoccus.
" 6. Glœocapsa.
" 7. Aphanocapsa.
" 8. Microcystis.
" 9. Clathrocystis.

Fig. 10. Cœlosphærium.
" 11. Merismopedia.
" 12. Nostoc.
" 13. Anabæna flos-aquæ.
" 14. Anabæna circinalis.

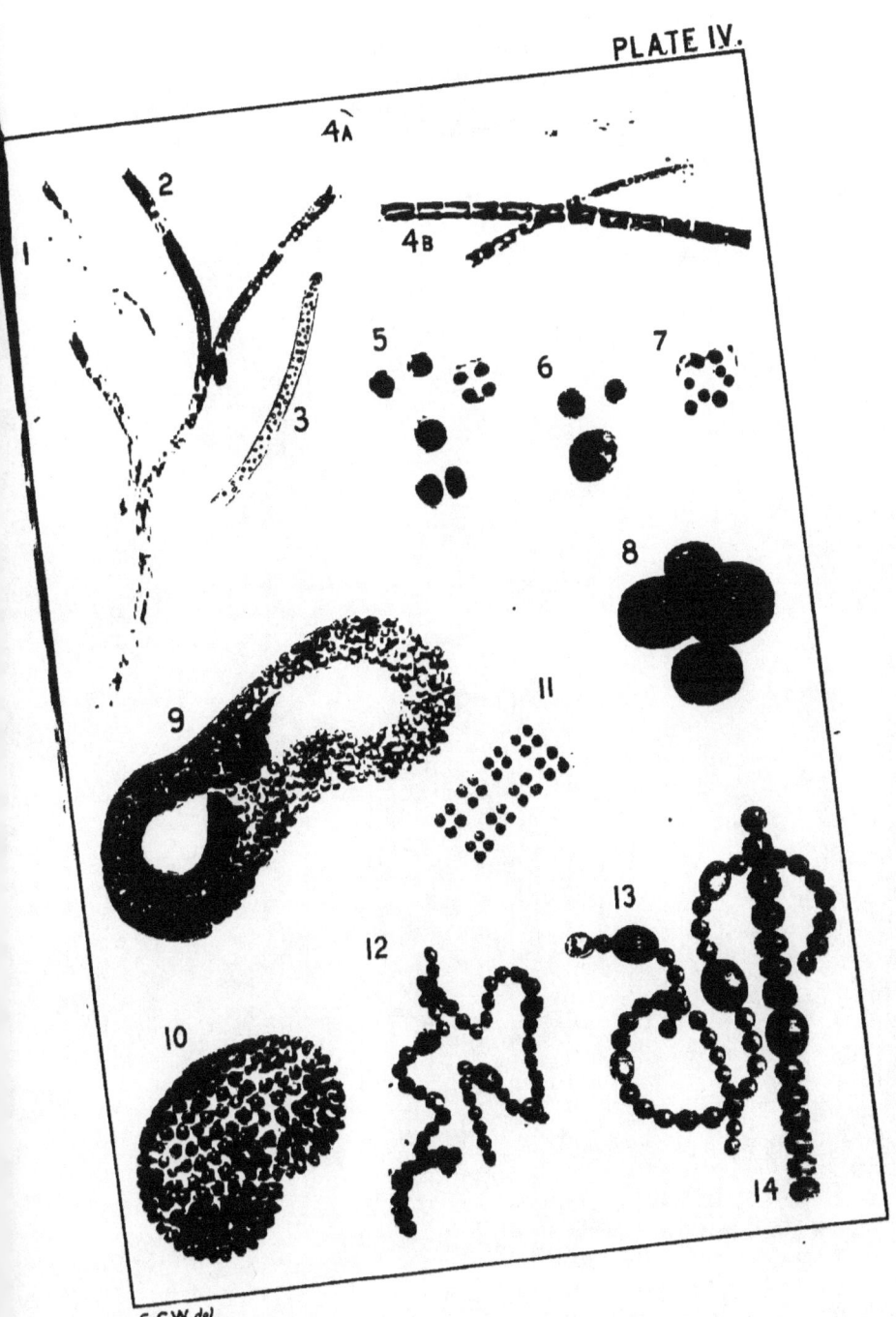

PLATE IV.

PLATE V.

CYANOPHYCEÆ. CHLOROPHYCEÆ.

PLATE V.

CYANOPHYCEÆ.

Magnification 500 diameters.

Fig. 1. Sphærozyga.
" 2. Cylindrospermum.
" 3. Aphanizomenon.
" 4. Oscillaria.
" 5. Lyngbya.

Fig. 6. Microcoleus.
" 7. Scytonema.
" 8. Sirosiphon.
" 9. Rivularia, a single filamen

CHLOROPHYCEÆ.

Magnification 500 diameters.

" 10. Glœocystis.
" 11. Palmella.

Fig. 12. Tetraspora.

PLATE V.

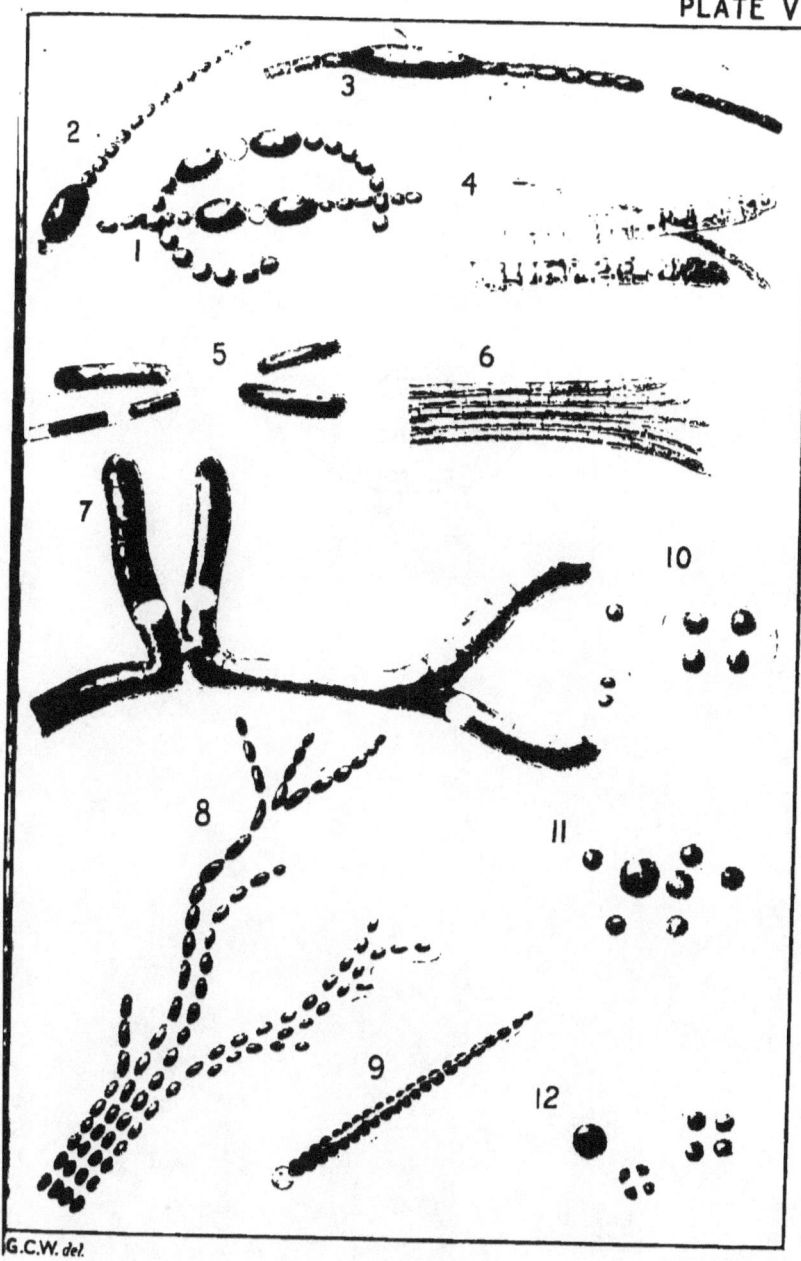

PLATE VI.
CHLOROPHYCEÆ.

PLATE VI.

CHLOROPHYCEÆ.

Magnification 500 diameters (except Fig. 9).

Fig. 1. Botryococcus.
" 2. Raphidium.
" 3. Dictyosphærium.
" 4. Nephrocytium.
" 5 Dimorphococcus.
" 6. Protococcus.

Fig. 7. Polyedrium.
" 8. Scenedesmus.
" 9. Hydrodictyon. ×250.
" 10. Ophiocytium.
" 11. Pediastrum.
" 12. Sorastrum.

PLATE VI.

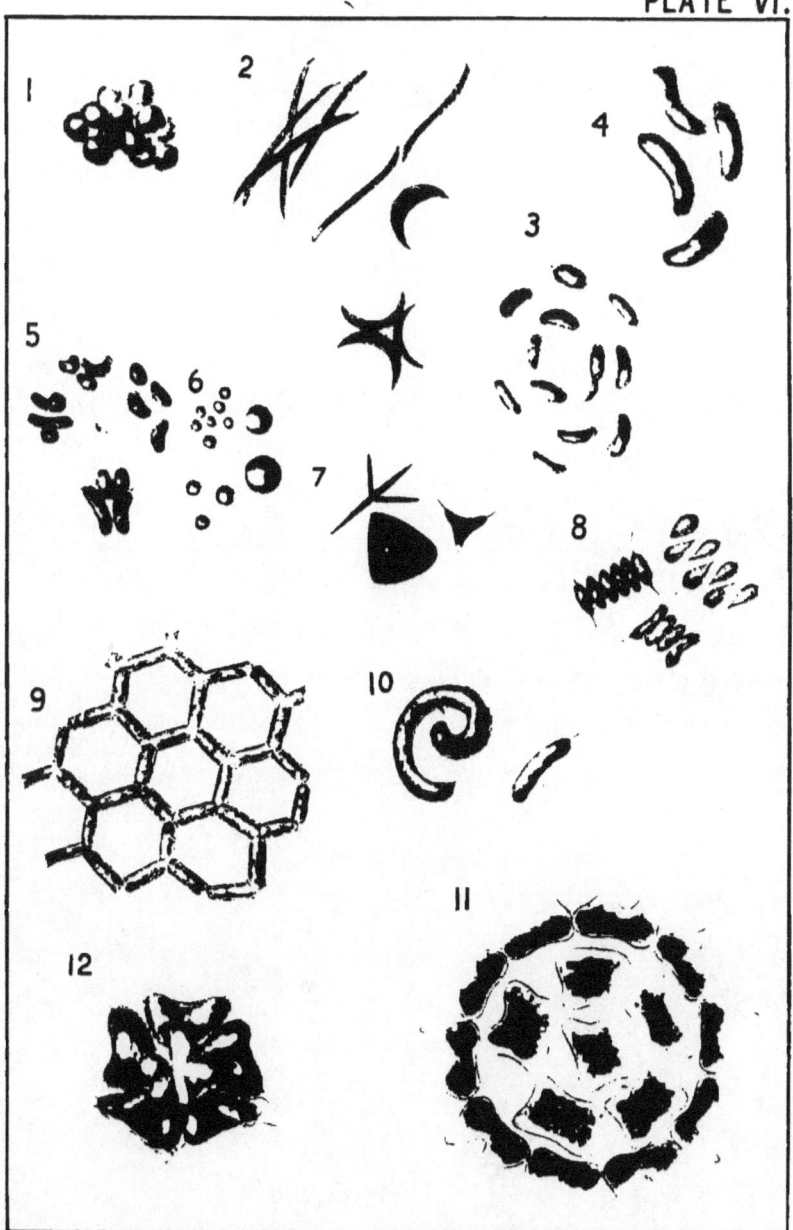

PLATE VII.

CHLOROPHYCEÆ.

PLATE VII.
CHLOROPHYCEÆ.

Fig. 1. Cœlastrum. × 500.
" 2. Staurogenia. × 500.
" 3. Volvox. × 100.
" 4. Eudorina. × 250.
" 5. Pandorina. × 250.
" 6. Gonium. *a*. top view.
  b, side view. × 500.
" 7. Penium. × 250.

Fig. 8. Closterium Dianæ. × 250.
" 9. Closterium Ehrenbergii. × 250.
" 10. Closterium subtile. × 250.
" 11. Docidium. × 250.
" 12. Cosmarium. × 250.
" 13. Tetmemorus. × 250.

PLATE VII.

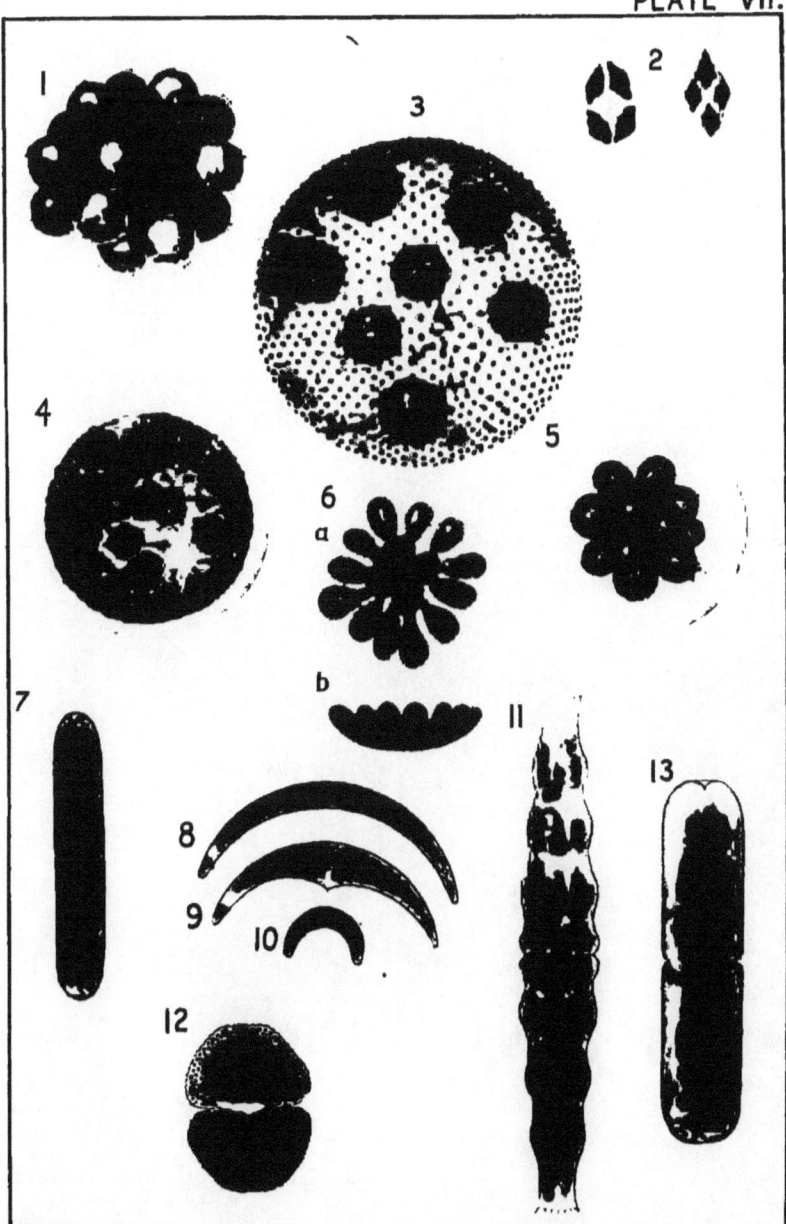

G.C.W. del.

PLATE VIII.

CHLOROPHYCEÆ.

PLATE VIII.
CHLOROPHYCEÆ.

Magnification 250 diameters.

Fig. A. Cosmarium, showing division.
Figs. B, C, D, E, and F. Cosmarium, showing conjugation, formation of zygospore and germination of the spore.

Fig. 1. Xanthidium armatum.
" 2. Xanthidium antilopæum. *a*, front view. *b*, lateral view. *c*, end view.
" 3. Arthrodesmus. *a*, front view. *b*, end view.
" 4. Euastrum. *a*, front view. *b*, lateral view.
" 5. Micrasterias.
" 6. Staurastrum magnum. *a*, front view. *b*, end view.
" 7. Staurastrum macrocerum. *a*, front view. *b*, end view.

PLATE VIII.

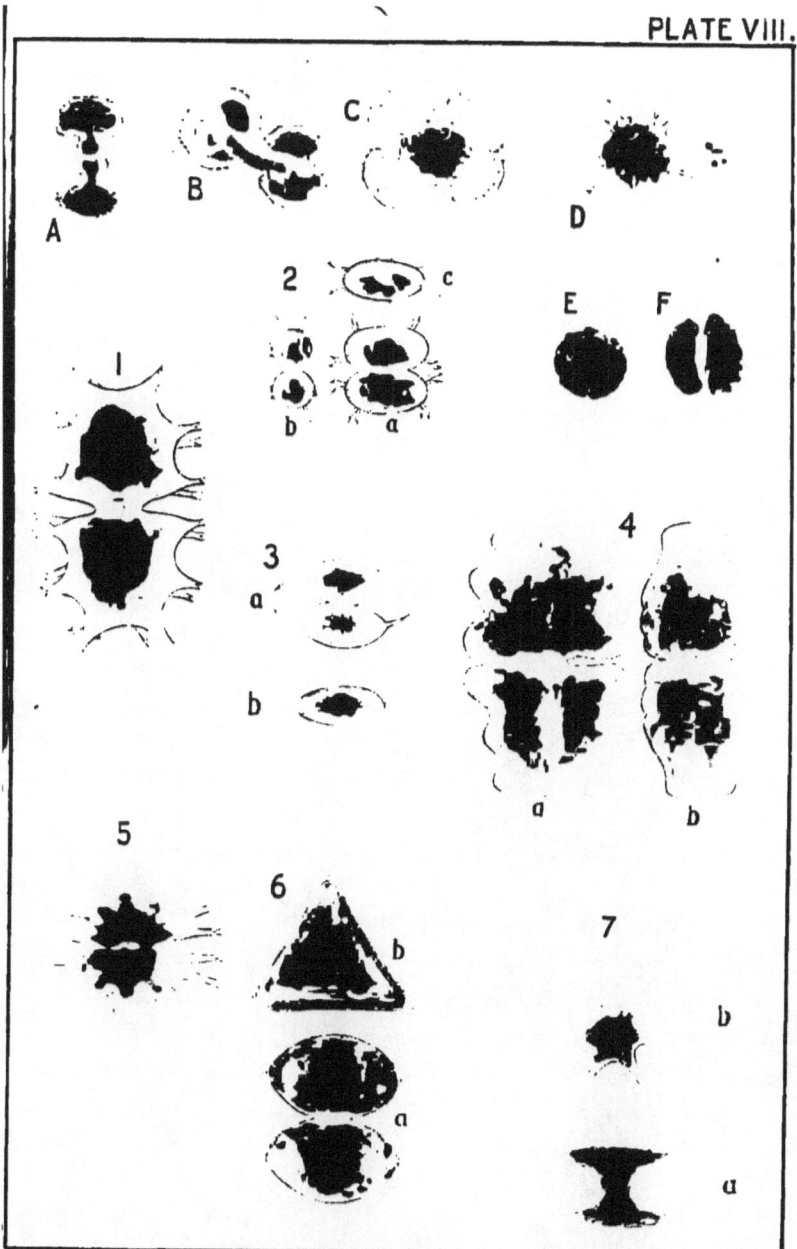

G.C.W. del.

PLATE IX.
CHLOROPHYCEÆ.

PLATE IX.

CHLOROPHYCEÆ.

Fig. 1. Hyalotheca. *a*, filament. *b*, end vi w. ×500.
" 2. Desmidium. *a*, filament. *b*, end view. ×500.
" 3. Sphærozosma. *a*, filament. *b*, end view. ×500.
" 4. Spirogyra. ×125.
" 5. Spirogyra, conjugated form, showing spores. ×125.
" 6. Zygnema. ×125.
" 7. Vaucheria. ×100.
" 8. Conferva. ×125.
" 9. Cladophora. ×75.
" 10. Ulothrix. ×125.

PLATE IX.

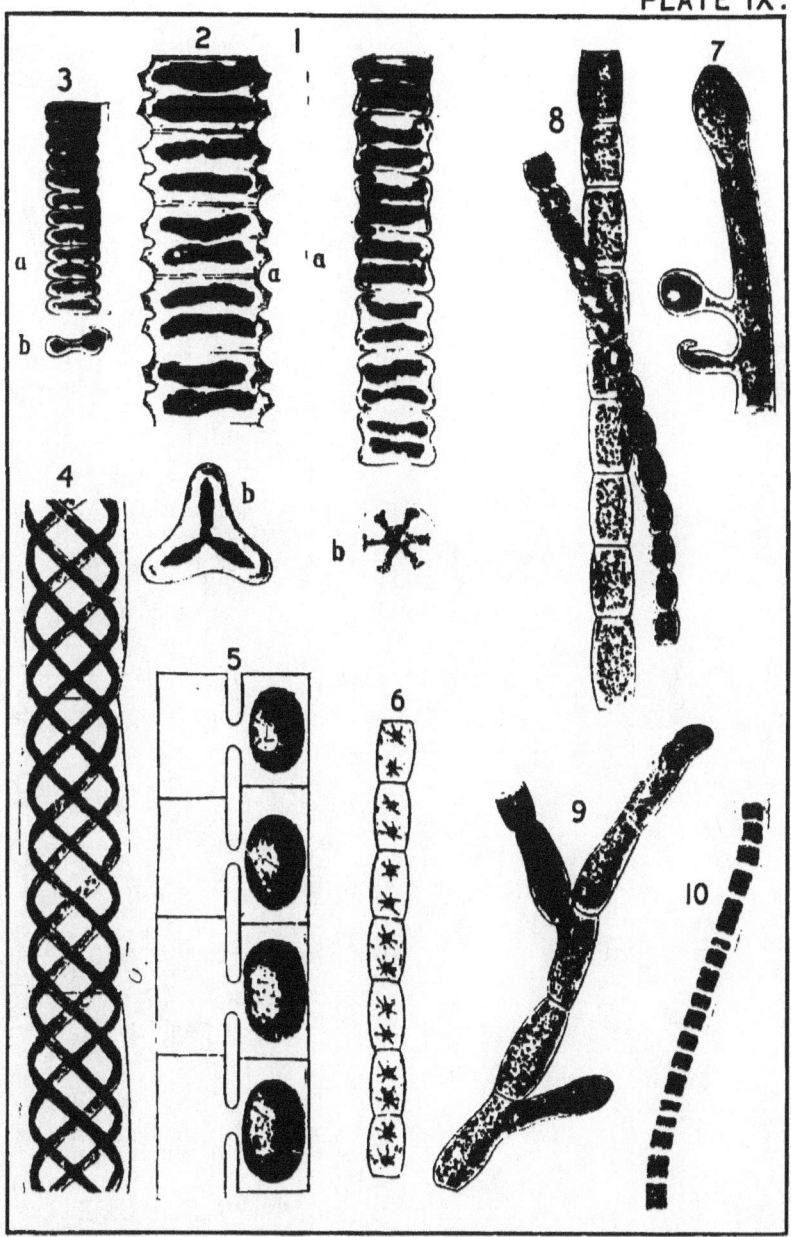

G.C.W. del.

PLATE X.
CHLOROPHYCEÆ. FUNGI.

PLATE X.

CHLOROPHYCEÆ.

Fig. 1. Stigeoclonium. ×125.   Fig. 3. Chætophora. ×125.
 " 2. Draparnaldia. ×125.

FUNGI.

Fig. 4. Saccharomyces. ×500.   Fig. 7. Aspergillus. ×250.
 " 5. Mold hyphæ. ×250.   " 8. Mucor. ×250.
 " 6. Penicillium. ×250.

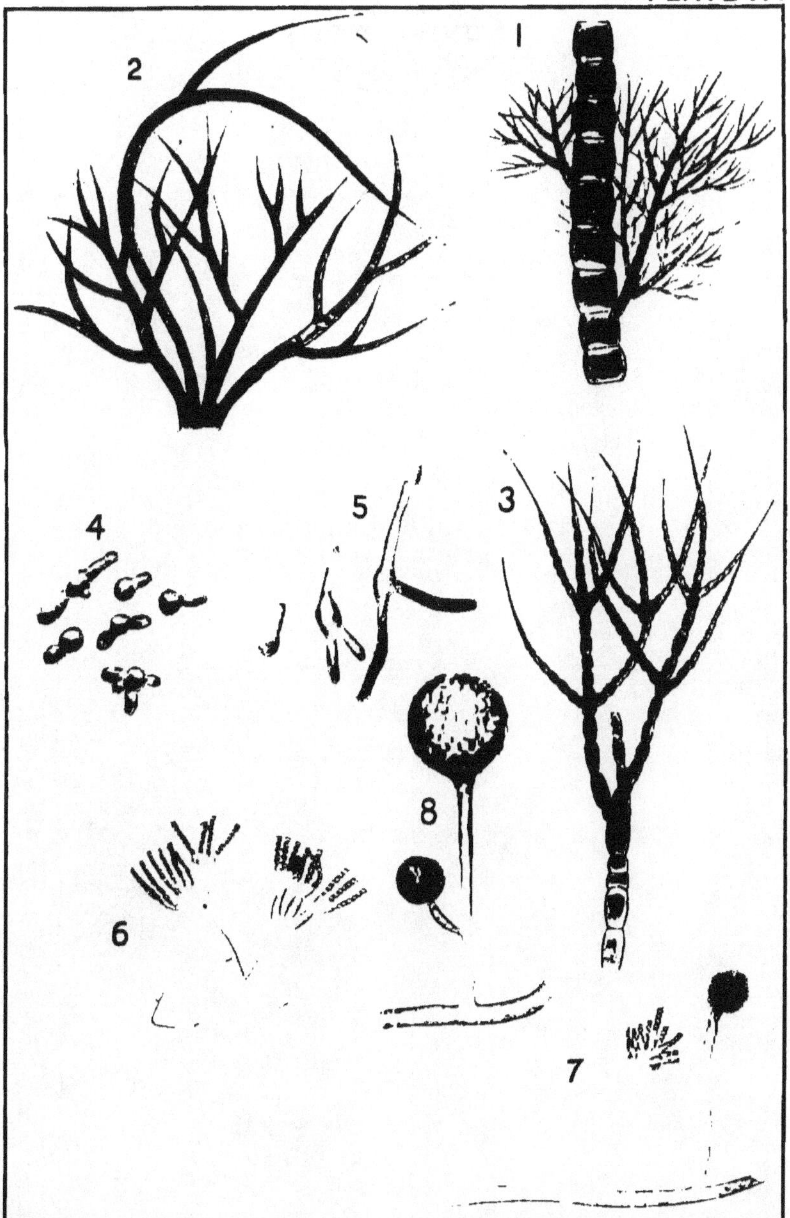

PLATE XI.
FUNGI. PROTOZOA.

PLATE XI.

FUNGI.

Fig. 1. Saprolegnia. × 250.   Fig. 3. Leptomitus. × 500.
" 2. Achlya. × 250.

PROTOZOA.

Fig. 4. Amœba. × 250.   Fig. 8. Euglypha. × 250.
" 5. Arcella, lateral view. × 250   " 9. Trinema. × 250.
" 6. Arcella, inferior view. × 250 " 10. Actinophrys. × 250.
" 7. Difflugia. × 250.

PLATE XI.

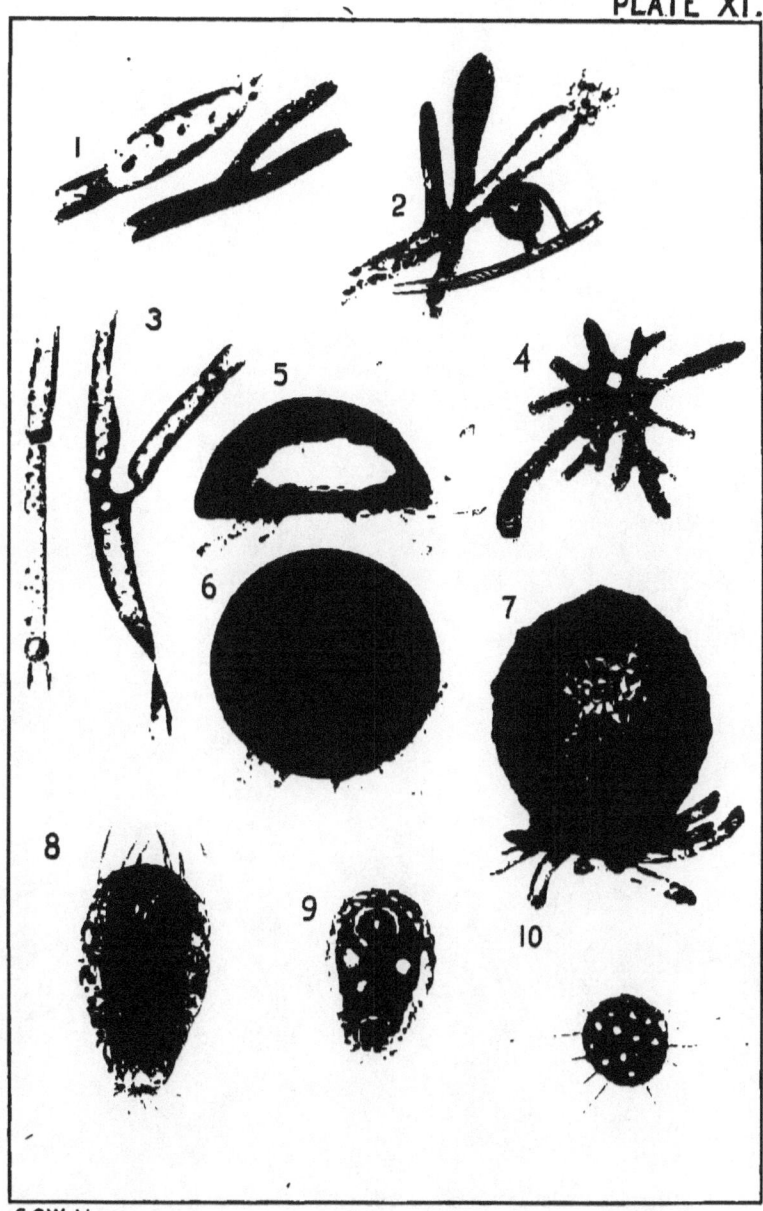

G.C.W. del.

PLATE XII.
PROTOZOA.

PLATE XII.
PROTOZOA.

Fig. 1. Cercomonas. × 500.
" 2. Monas. × 500.
" 3. Anthophysa. × 500.
" 4. Cœlomonas. × 500.
" 5. Raphidomonas. × 500.
" 6. Euglena. × 500.
" 7. Trachelomonas. × 500.

Fig. 8. Phacus. × 500.
" 9. Synura. × 500.
" 10. Uvella. × 500.
" 11. Syncrypta. × 500.
" 12. Uroglena. × 250.
" 13. Uroglena; showing division of the monads. × 1000.

PLATE XII.

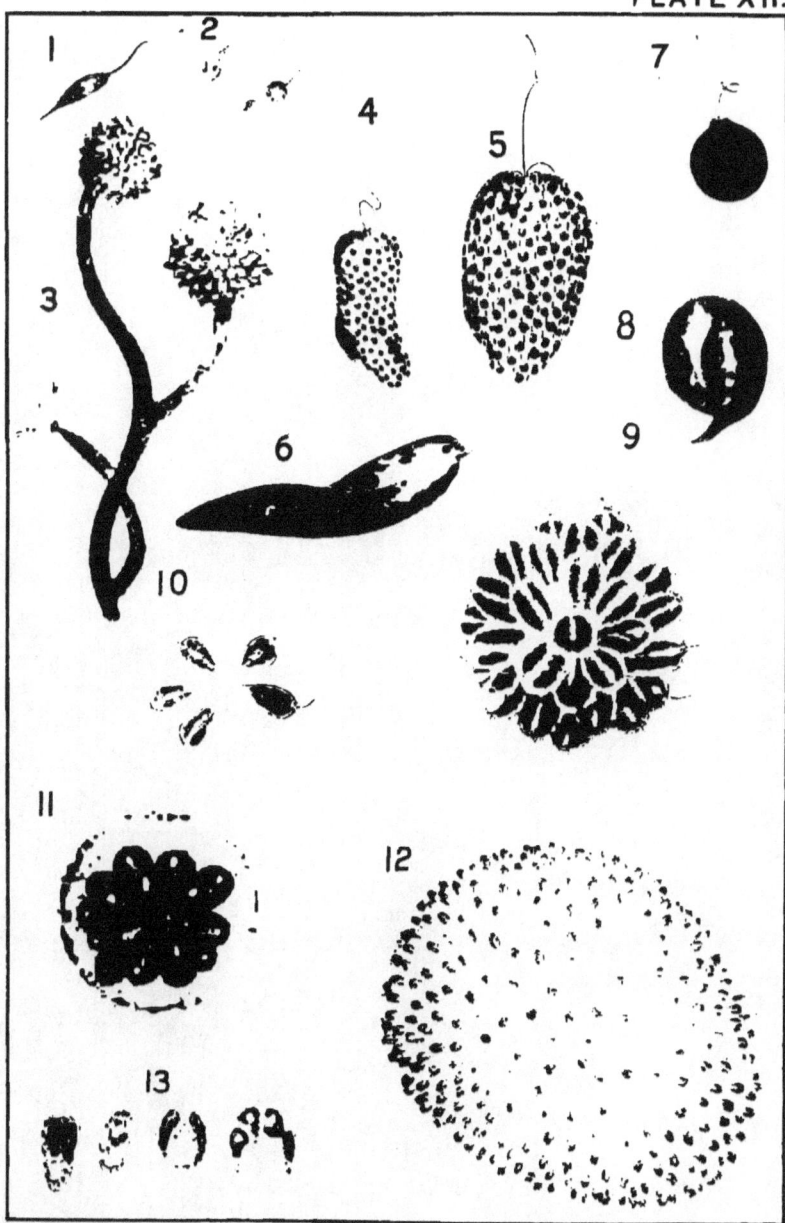

G.C.W. del.

PLATE XIII.

PROTOZOA.

PLATE XIII.

PROTOZOA

Fig. 1. Dinobryon. × 500.
" 2. Cryptomonas. × 500.
" 3. Mallomonas. × 500.
" 4. Chlamydomonas. × 500.
" 5. Peridinium. × 500.
   6. Ceratium. × 250.

Fig. 7. Glenodinium. × 500.
" 8. Euplotes. × 250.
" 9. Halteria. × 500.
" 10. Vorticella. × 250.
" 11. Epistylis. × 250.
" 12. Tintinnus. × 250.

PLATE XIII.

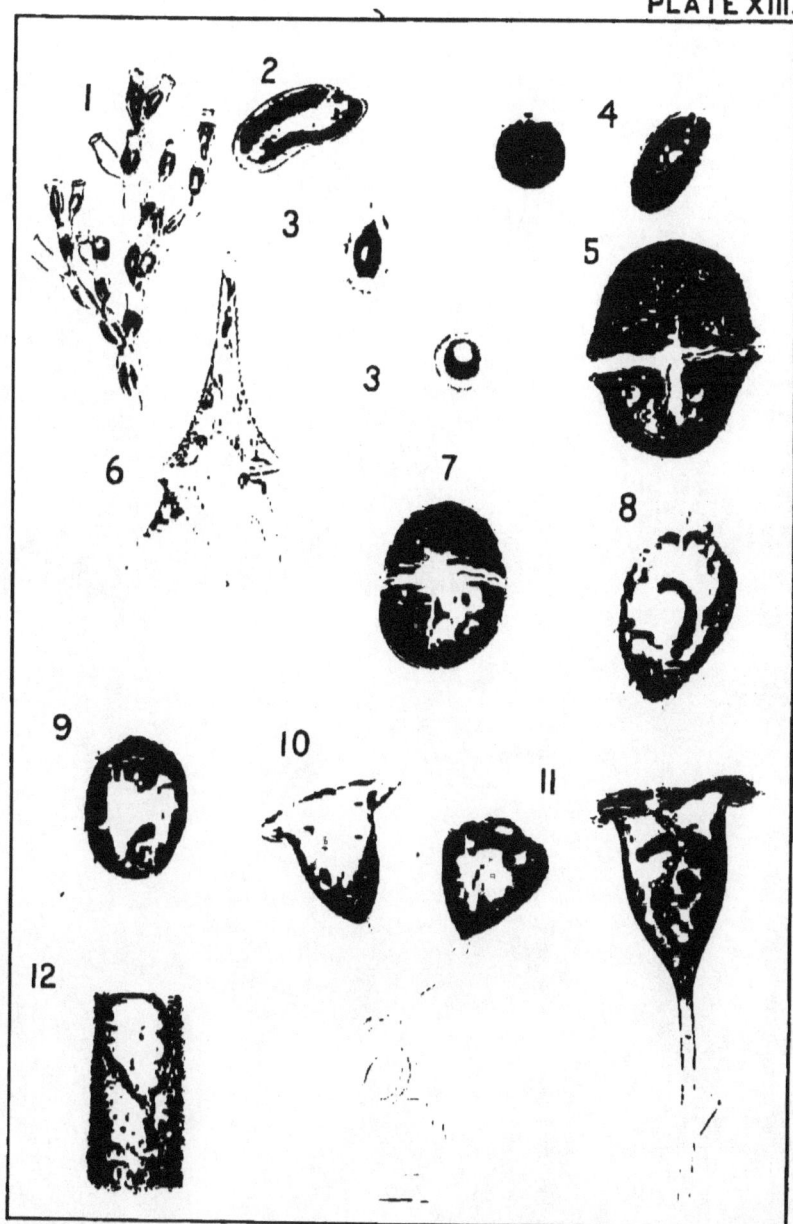

G.C.W. del.

PLATE XIV.
PROTOZOA.

PLATE XIV.
PROTOZOA.

Fig. 1. Codonella. ×500.
" 2. Stentor. ×50.
" 3. Bursaria. ×100.
" 4. Paramœcium. ×250.
" 5. Nassula. ×250.

Fig. 6. Coleps. ×500.
" 7. Enchelys. ×500.
" 8. Trachelocerca. ×500.
" 9. Pleuronema. ×500.

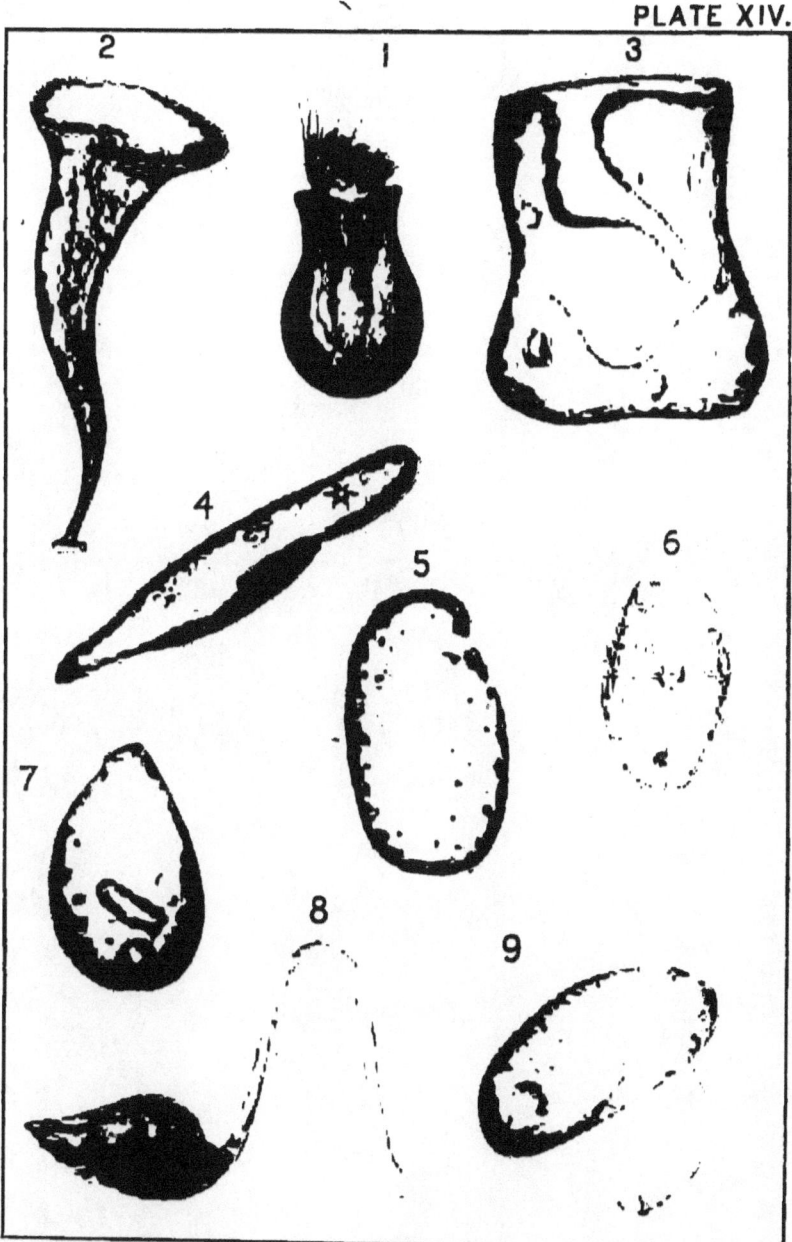

PLATE XIV.

PLATE XV.
PROTOZOA. ROTIFERA.

PLATE XV.

PROTOZOA.

Fig. 1. Colpidium. ×500. Fig. 2. Acineta. ×500.

ROTIFERA.

Fig. 3. Floscularia. ×25.
" 4. Melicerta. ×25.
" 5. Conochilus. ×100.

Fig. 6. Rotifer. ×100.
" 7. Microcodon. ×150.
" 8. Asplanchna. ×150.

PLATE XV

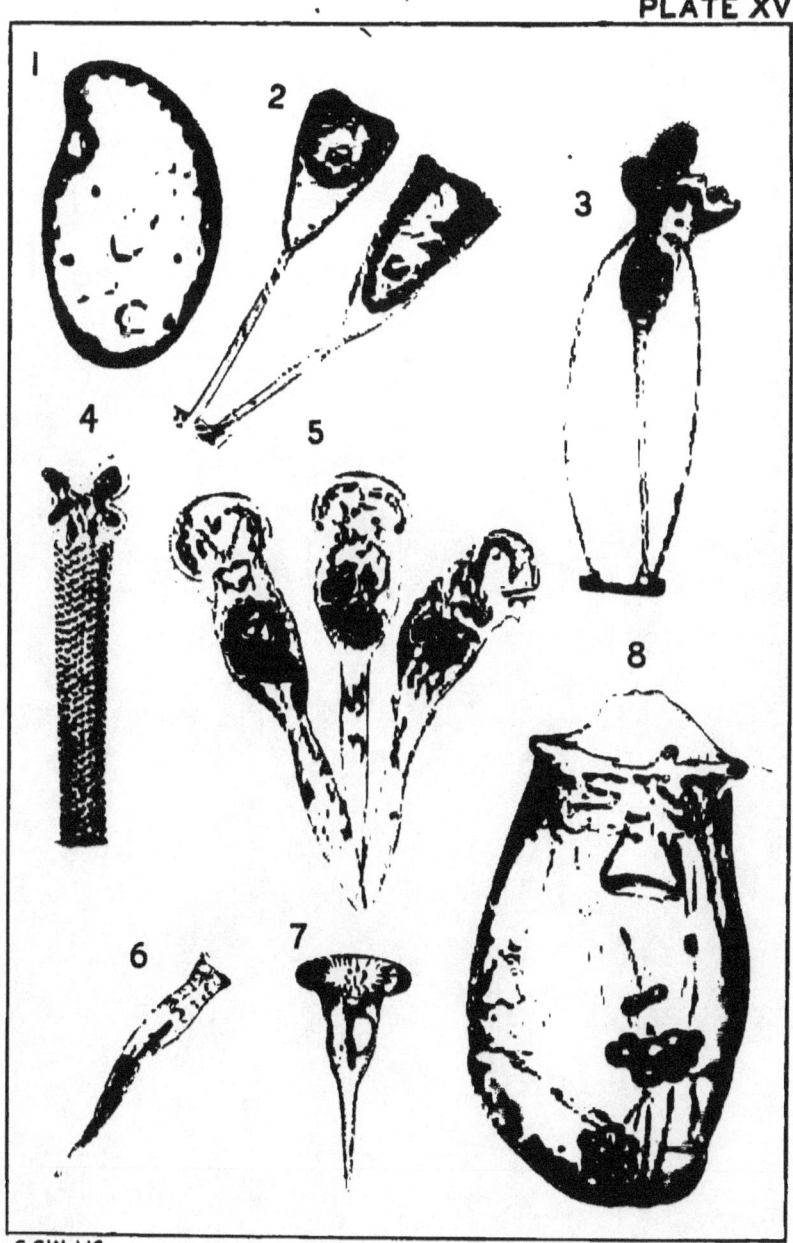

G.C.W. del.

PLATE XVI.

ROTIFERA.

PLATE XVI.

Figs. A to E. Diagrams of Trochal Disc. (After Bourne.)
A, Microcodon. B, Stephanoceros. C, Hypothetical form intermediate between Microcodon and Philodina. D Philodina. E, Brachionus.

Figs. F to I. Diagrams showing Structure of the Foot. (After Hudson and Gosse.)
F, Rhizotic foot (Floscularia). G, Rhizotic foot (Melicerta) H, Bdelloidic foot (Rotifer). I, Scirtopodic foot (Pedalion).

Figs. J to P. Diagrams showing Forms of Trophi. (After Hudson and Gosse.)
J, Malleate. K, Sub-malleate. L, Forcipitate. M, Incudate. N, Uncinate. O, Ramate. P, Malleo ramate.

ROTIFERA.

Fig. 1. Synchæta. ×100.
" 2. Polyarthra. ×200.
" 3. Triarthra. ×150.

Fig. 4. Diglena ×150.
" 5. Mastigocerca. ×150

PLATE XVI.

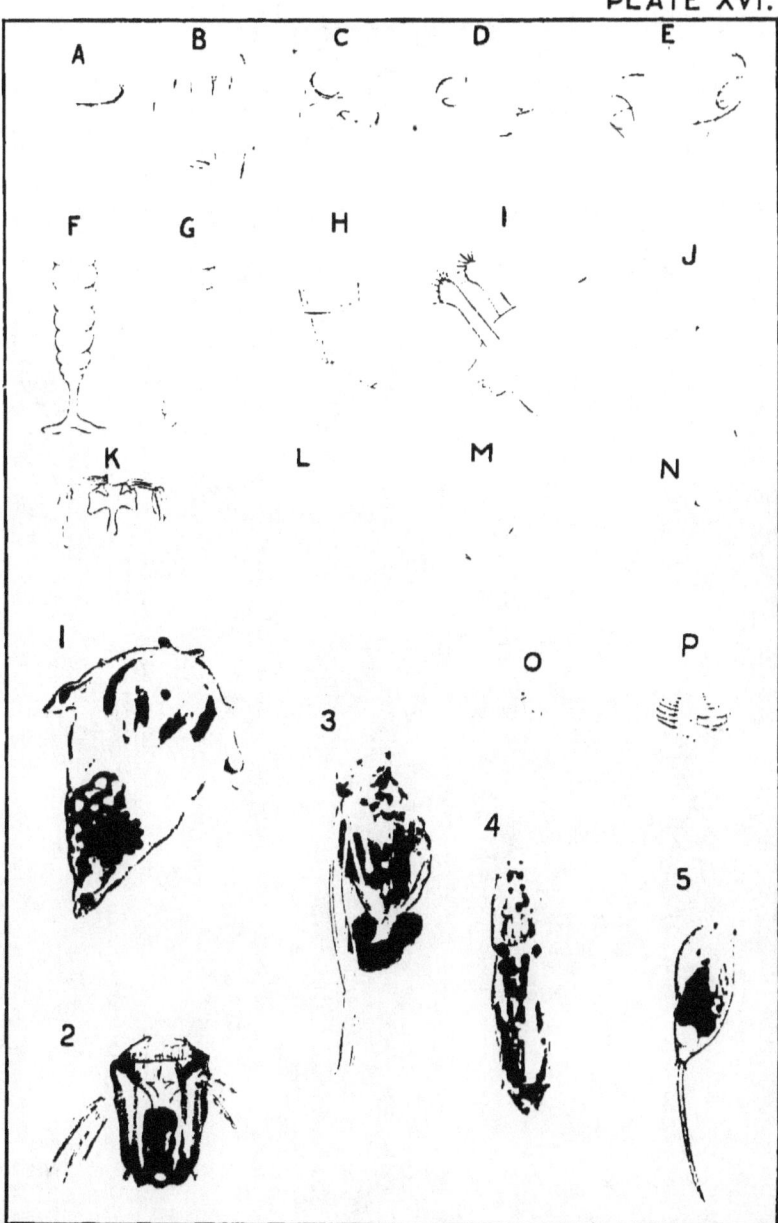

G.C.W. del.

PLATE XVII.
ROTIFERA. CRUSTACEA.

PLATE XVII.

ROTIFERA.

Fig. 1. Brachionus. ×200.
" 2. Anuræa cochlearis. *A*, dorsal view. *B*, side view. ×150.
Fig. 3. Anuræa aculeata. ×150.
" 4. Notholca. ×200.

CRUSTACEA.

Fig. 5. Cyclops. ×25.
" 6. Diaptomus. ×25.
" 7. Canthocamptus. ×25.
Fig. 8. Cypris. ×25.
" 9. Daphnia. ×25.
" 10. Bosmina. ×25.

PLATE XVII.

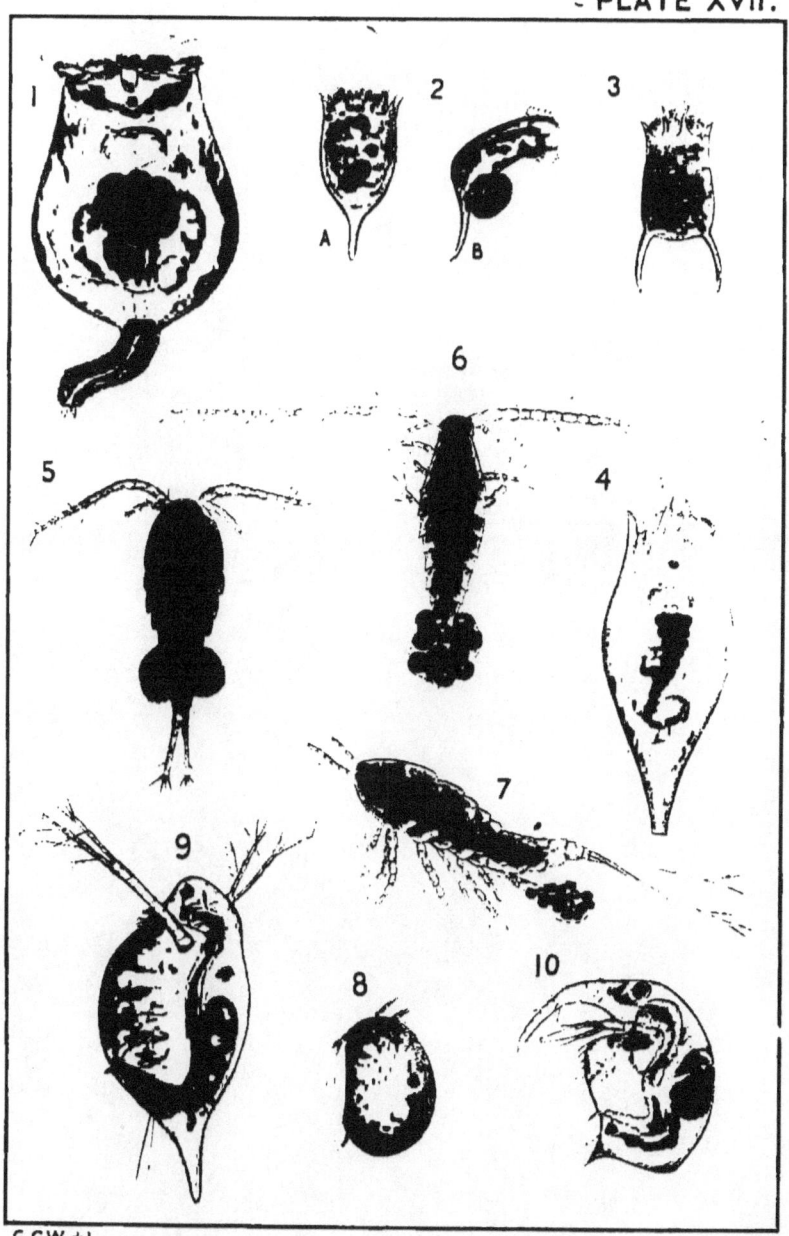

PLATE XVIII.
CRUSTACEA. BRYOZOA. SPONGIDÆ.

PLATE XVIII.

CRUSTACEA.

Fig. 1. Sida. ×25.
" 2. Chydorus. ×25.

Fig. 3. Branchipus. ×2.

BRYOZOA.

Fig. 4. Fredericella. ×5.
" 5. Paludicella. ×5.

Fig. 6. Statoblast of Plumatella. ×25.
" 7. Statoblast of Pectinatella. ×25.

SPONGIDÆ.

Fig. 8. Spongilla. ×1.
" 9. Sponge spicules (skeleton spicules). ×150.

PLATE XVIII.

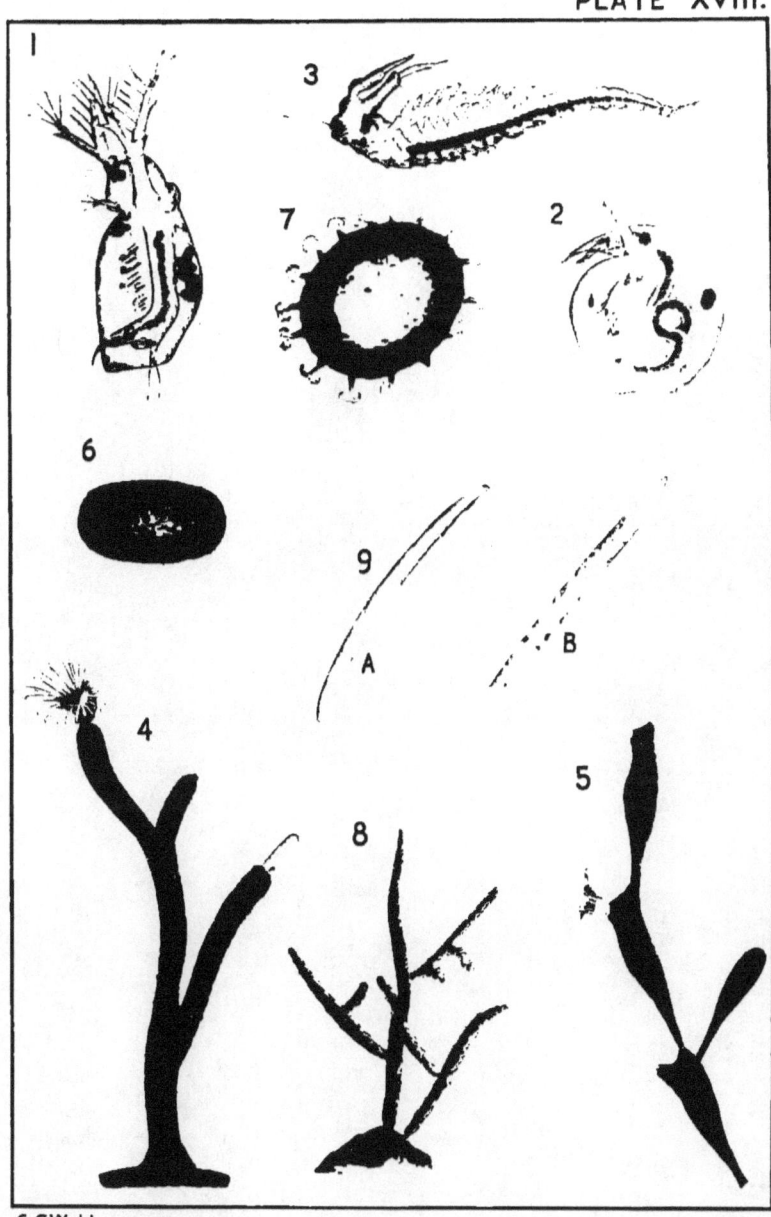

PLATE XIX.
MISCELLANEOUS.

PLATE XIX.
MISCELLANEOUS.

Fig. 1. Anguillula. × 100.
" 2. Nais. × 10.
" 3. Chætonotus. × 250.
" 4. Macrobiotus. × 250.
" 5. Acarina. × 25.
" 6. Hydra. × 25.

Fig. 7. Batrachospermum. × 100.
" 8. Chara. × 75.
" 9. Anacharis. × 1.
" 10. Ceratophyllum. × 1.
" 11. Potamogeton. × 1.
" 12. Lemna. × 1.

PLATE XIX.

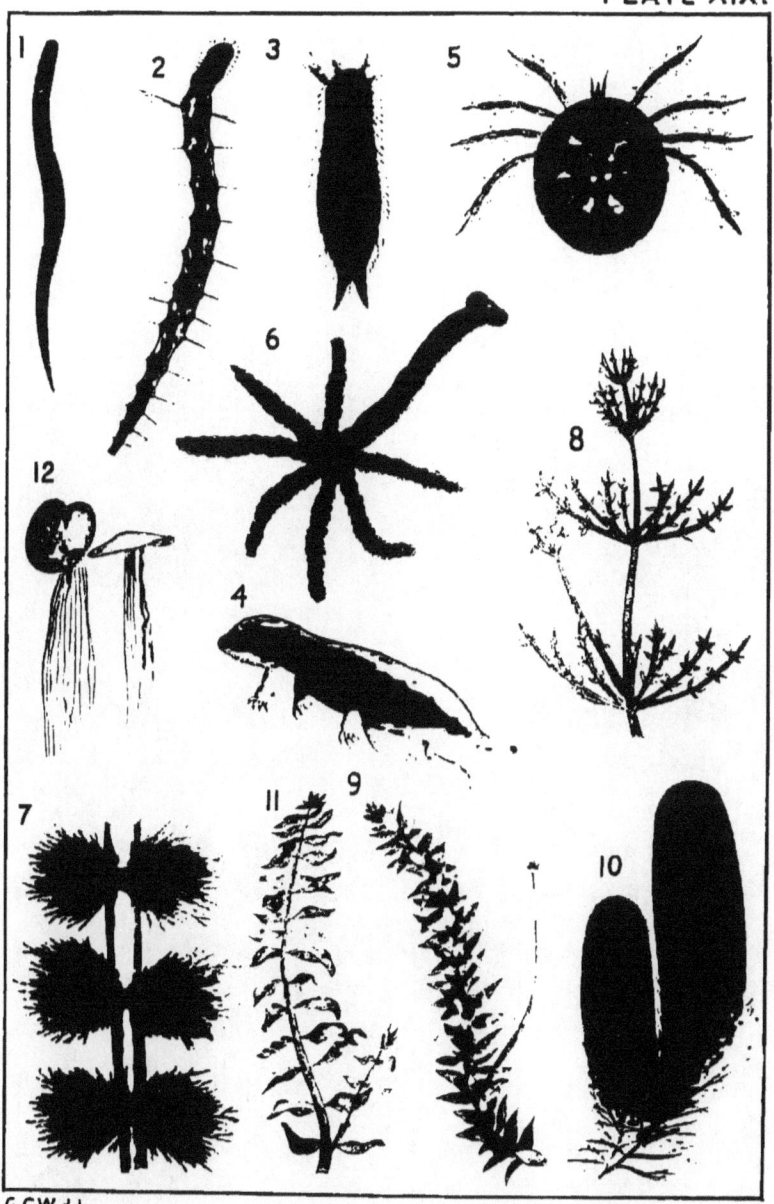

www.ingramcontent.com/pod-product-compliance
Lightning Source LLC
Chambersburg PA
CBHW030342230426
43664CB00007BA/506